VALUING HISTORIC ENVIRONMENTS

Heritage, Culture and Identity

Series Editor: Brian Graham,
School of Environmental Sciences, University of Ulster, UK

Valuing Historic Environments

Edited by

LISANNE GIBSON
University of Leicester, UK

and

JOHN PENDLEBURY
Newcastle University, UK

LONDON AND NEW YORK

First published 2009 by Ashgate Publishing

2 Park Square, Milton Park, Abingdon, Oxon OX14 4RN
711 Third Avenue, New York, NY 10017, USA

Routledge is an imprint of the Taylor & Francis Group, an informa business

First issued in paperback 2016

British Library Cataloguing in Publication Data
Valuing historic environments. - (Heritage, culture and
    identity)
    1. Cultural property - Protection
    I. Gibson, Lisanne II. Pendlebury, John
    363.6'9

Library of Congress Cataloging-in-Publication Data
Valuing historic environments / [edited] by Lisanne Gibson and John Pendlebury.
    p. cm. -- (Heritage, culture and identity)
    Includes bibliographical references and index.
    ISBN 978-0-7546-7424-5 -- ISBN 978-0-7546-9043-6 (ebook)
    1. Cultural property--Protection. 2. Antiquities--Collection and preservation. 3.
Historic sites--Conservation and restoration. 4. Landscape protection. I. Gibson,
Lisanne. II. Pendlebury, John R., 1963-

CC135.V35 2009
363.6'9--dc22

                                                                              2008054350

ISBN 13: 978-0-7546-7424-5 (hbk)
ISBN 13: 978-1-138-25743-6 (pbk)

Contents

PART III THE HERITAGE OF HOUSING

List of Figures

List of Tables

List of Contributors

Tracey Avery, Heritage Consultant, Melbourne and formerly The National Trust of Australia (Victoria)

Professor Peter Borsay, Department of History and Welsh History, Aberystwyth University, UK

Dr Lisanne Gibson, Department of Museum Studies, University of Leicester, UK

Rose Gilroy, School of Architecture, Planning and Landscape, Newcastle University, UK

Peter Howard, Visiting Professor of Cultural Landscapes, School of Conservation Sciences, Bournemouth University, UK

Professor David Lowenthal, Department of Geography, University College London, UK

Peter Malpass, Professor of Housing and Urban Studies, School of the Built and Natural Environment, University of the West of England, Bristol, UK

John Pendlebury, School of Architecture, Planning and Landscape, Newcastle University, UK

Dr John Schofield, Characterization Team, English Heritage, UK

Tim Townshend, School of Architecture, Planning and Landscape, Newcastle University, UK

Dr Laurajane Smith, Department of Archaeology, University of York, UK

Professor John K. Walton, Institute of Northern Studies, Leeds Metropolitan University, UK

Professor Jason Wood, Institute of Northern Studies, Leeds Metropolitan University, UK

Acknowledgements

This book emerged out of a research cluster, organized by the editors of this text, titled 'Valuing Historic Environments: Concepts, Instrumentalizations and Effects'. The key activity of the network was a series of well attended and lively workshops with contributions from academia and heritage practice. All but two of the chapters in this book started life as papers commissioned for those events. We are indebted not only to those who developed excellent presentations for those workshops but to all those who took part and in particular those individuals who formed the core of our cluster.

The 'Valuing Historic Environments' research cluster was funded through a special cross-council funding programme called 'Preserving Our Past' and we would like to acknowledge the support of the Engineering and Physical Sciences Research Council, the Arts and Humanities Research Council, the Economic and Social Research Council, the Natural Environment Research Council and English Heritage in providing financial support for the programme and our cluster.

Finally, we would like to thank Anna Woodham who helped us both with running the cluster events and compiling this book.

Introduction:
Valuing Historic Environments

Lisanne Gibson and John Pendlebury

Introduction

There is a contemporary imperative to consider different cultural, historical and social values as equal. If we accept this and take it take seriously, how do we overcome the practical issues this presents for the management as heritage of historic environments and cultural landscapes? This single question is probably the most significant issue facing contemporary heritage management and policy. It is this challenge to heritage management that this rich compilation of essays on a wide diversity of environments (physical and institutional) and landscapes aims to address. If the socially democratic context for our contemporary understanding of value is one of pluralization, involving the validation of multiple conceptions of value, what does this mean for acts of preservation which are, by their very nature, based on processes which involve the fixing of meaning and value?

All of the authors contributing to this collection proceed on the basis that concepts of cultural, historical, or social value are culturally and historically constructed. This theoretical orientation posits most crucially that value is not an *intrinsic* quality but rather the fabric, object or environment is the bearer of an externally imposed culturally and historically specific meaning, that attracts a value status depending on the dominant frameworks of value of the time and place. Such an orientation has consequences both for the assessment of significance and the heritage management of a building, object or environment.

That concepts of value are constructed has become the dominant theoretical approach across the humanities and social sciences. In a democratic society, it is argued, definitions of value cannot be singular but must allow for plural interpretations and meaning. In the heritage field this 'cultural turn' has led to a questioning of what constitutes value.[1] This has resulted in erosion of the previously dominant notion of value which understood it as intrinsic to the object or environment and able to be revealed by the correct processes of investigation which could be conducted only by a limited body of experts.

---

1   See for instance, Avrami, E., et al., *Values and Heritage Conservation*, (Los Angeles: The Getty Conservation Institute, 2000) and De La Torre, M., *Assessing the Values of Cultural Heritage*, (Los Angeles: The Getty Conservation Institute, 2002).

In part this understanding of the validity of multiple concepts of value emanates from the recognition that acts of preservation have cultural, economic, political, and social consequences. The preservation of an object or environment is an assertion of its importance and therefore the culture or history associated with it. Such artefacts or environments and the meanings they represent are often integrally tied to the identity formations of particular groups or communities. This can have both positive and negative effects.[2] On the one hand, if singular understandings of value are prioritized and supported over others the consequences are the support of one group's culture and not another's.[3] As Lisanne Gibson and Laurajane Smith discuss in their contributions to this book the support of particular cultural forms may have political, social and even economic effects through the support of particular systems of 'cultural capital'. On the other hand, the recognition of the connections between identities and objects or environments has also led to a multiplication of programmes, including heritage programmes, which in recognizing the validity of plural value frameworks seek to enable and empower a diversity of communities and individuals through actively supporting and valuing their stories, objects and places.[4] Thus, while cultural support (funding or preservation) can be articulated to socially democratic discourses we must nevertheless remain aware that cultural support involves choices which have cultural, economic, political and social ramifications.[5]

The burgeoning of publicly funded cultural support, including for heritage, since the 1970s and the recognition that there are cultural, economic, political and social effects of cultural support has led to a requirement that cultural programmes must affect social and public policy outcomes, such as regeneration.[6] This shift has put increasing pressure on cultural programmes to respond to (shifting) Government objectives rather than responding to logics which are defined internally to the field.[7] Recently the cultural policy, museum and heritage studies literatures have

---

2 L. Gibson and J. Besley, *Monumental Queensland: Signposts on a Cultural Landscape*, (Brisbane: University of Queensland Press, 2004), 8–15, see also Gibson this volume and Smith this volume.

3 P. Bourdieu, *Distinction: A Social Critique of the Judgement Of Taste*, (Oxford: Routledge, 1984); T. Bennett, and E.B. Silva, 'Cultural Capital and Inequality: Policy Issues and Contexts', *Cultural Trends*, 15, 2/3 (2006): 87–106; and T. Bennett and M. Savage, 'Introduction: Cultural Capital and Cultural Policy', *Cultural Trends* 13, 2, (2004): 7–14.

4 P. Davis, *Ecomuseums: A Sense of Place*, (Leicester: Leicester University Press, 1999); A. Newman, F. McLean, and G. Urquhart, 'Museums and the active citizen: tackling the problems of social exclusion'. *Citizenship Studies*, 9, 1 (2005): 41–57.

5 L. Gibson, *The Uses of Art: Constructing Australian Identities*, (Brisbane: University of Queensland Press, 2001); J. Pendlebury et al., 'The Conservation of English Cultural Built Heritage: A Force for Social Inclusion?', *International Journal for Heritage Studies*, 10, 1 (2004): 11–32.

6 L. Gibson, *The Uses of Art: Constructing Australian Identities*, op. cit.; L. Gibson, 'In defence of Instrumentality', *Cultural Trends*, 17, 4, (2008): 247–257.

7 J. Pendlebury, *Conservation and the Age of Consensus*, (London: Routledge, 2009); I. Strange and D. Whitney, 'The changing roles and purposes of heritage conservation in

contained a great deal of discussion of this so called 'instrumentalization' of cultural institutions and programmes which is described by this literature as emerging over the last thirty or so years. This perception of culture's co called 'instrumentalization' seems to be widespread and is primarily perceived as a 'threat'.[8] However, in these deconstructions, primarily aimed at the poor impact studies and overblown claims made for the arts and heritage, there is little to guide us towards a way of thinking which takes seriously the practical challenges for cultural and heritage management and policy. Our purpose is querying the utility of such critiques is not to propose a simple acceptance of the status quo, far from it. Rather, it is to identify that one of the problem's with the attack on the 'instrumentalization' of culture is that it leaves the field open for a return to the kinds of elite, exclusionary policies which have characterized cultural administration in the past, and in many cases still do. As Mark O'Neill, Head of Museums and Galleries for Glasgow City Council, concludes in his critique of John Holden's *Capturing Cultural Value*,

> Targets and measurements can be refined, but what can be done about the profound sense amongst... groups of entitlement – entitlement to having their cultural recreations funded without being troubled by the values of a wider society based on democracy, accountability, equity and fairness?[9]

Thus, in addition to more 'abstract' considerations of the pluralization of value in relation to heritage, we are also concerned to focus discussion on the practical and grounded applications, contexts and outcomes of heritage. This is not least because we believe that one of the key issues for contemporary heritage management is how to ensure that it is democratic while at the same time ensuring that the heritage profession does not become a 'Government poodle'?[10]

In introducing this book we would like to take the opportunity to examine more closely some of the coordinates we have sketched out here. What are some of the key historical and cultural constellations for contemporary typologies of value? How has the post-modern pluralization of value been instrumentalized in heritage management?

---

the UK', *Planning Practice and Research*, 18, 2–3 (2003): 219–229; C. West and C. Smith, '"We are not a Government poodle": Museums and social inclusion under New Labour', *International Journal of Cultural Policy*, 11, 3 (2005): 275–288.

8 E. Belfiore, 'Art as a means of alleviating social exclusion: Does it really work? A critique of instrumental cultural policies and social impact studies in the UK', *International Journal of Cultural Policy*, 8, 1 (2002): 91–106; J. Holden, *Capturing Cultural Value: How Culture has Become a Tool of Government Policy*, (Demos: London, 2004); S. Selwood, 'Measuring Culture', *Spiked Culture*, http://www.spiked-online.com/Printable/00000006DBAF.htm, (2002), accessed 01/09/2003.

9 M. O'Neill, 'Commentaries: John Holden's *Capturing Cultural Value: How Culture has Become a Tool of Government Policy*', *Cultural Trends*, 14(1), 53 (2005): 124.

10 C. West and C. Smith, '"We are not a Government poodle": Museums and social inclusion under New Labour', op. cit.

Some of the fundamental premises of the debates explored in this book – the cultural and historical specificity of heritage, the pluralization of value – can be briefly illustrated through examination of the case of perhaps the most iconic British monument of all, Stonehenge. As writers as diverse as the archaeologist Barbara Bender,[11] the architectural theorist Andrew Ballantyne[12] and the historian Andy Worthington[13] have attested the monument has a long history of being interpreted in multiple and conflicting ways; interpretations that are often linked to contemporary purposes and needs. For example, Bender describes how medieval religious attitudes shifted, hardening to become antagonistic to these pre-Christian relics. Subsequently, in the seventeenth and eighteenth centuries, the (now) well-known works of Inigo Jones, John Aubrey and William Stukeley influenced by contemporary discourses of nationalism and their own cultural pre-dispositions, interpreted Stonehenge as variously a Roman and Druidic monument. As Jacquetta Hawkes wrote in 1967, 'Every age has the Stonehenge it deserves – or desires'.[14] The site remains enigmatic; a Foucaldian 'heterotopia', a place that is different from itself, as 'people immersed in different cultures come and visit the place they continue to occupy and experience different spaces – Druid space, Merlin space etc.'[15] These different interpretations of Stonehenge fit within a wider ideological and political discourse about whose values should prevail, with competing and conflicting positions. This was most dramatically evident in the 1980s when a right-wing government (in conjunction with heritage agencies) deployed the sorts of repressive techniques concurrently used in the Miners' Strike, against the heterogeneous, if typically non-conformist, free-festival goers at the so-called Battle of the Beanfield in 1985. State will prevailed and although attitudes have subsequently softened with, for example, limited admission of druids to Stonehenge on the summer solstice, Stonehenge remains a highly controlled and commodified environment. Nevertheless, even within the various agencies responsible for the management of Stonehenge conflicts of value are evident – proposals for putting a section of the A303 adjacent to Stonehenge underground and for new visitor facilities have been phenomenally contentious and disputed for over a decade and were dramatically abandoned in late 2007. There is general consensus that the existing visitor facilities are inadequate and poorly sited and that the nearby roads, the A303 long-distance trunk-road and the more local A344 are hugely intrusive. However, it has proved impossible to generate consensus on how to resolve this, with disagreements over siting and how the wider landscape is

---

11   B. Bender, *Stonehenge: Making Space*, (Oxford: Berg, 1998).

12   A. Ballantyne, 'Misprisions of Stonehenge' in *Architecture as Experience: Radical Change in Spatial Practice* edited by D. Arnold and A. Ballantyne, 11–35, (London: Routledge: 2004).

13   A. Worthington, *Stonehenge: Celebration and Subversion*, (Loughborough: Alternative Albion, 2004).

14   Cited in B. Bender, *Stonehenge: Making Space*, 114, op. cit.

15   A. Ballantyne, 'Misprisions of Stonehenge', 28, op. cit.

understood, issues over how much financial resource the Government will devote to improvements and wider complications, such as the large Ministry of Defence operation in the area. Though major road proposals have been abandoned there remains a commitment to improving visitor facilities by the time of the London Olympics in 2012, with a consultation exercise over the form this might take being undertaken in summer 2008.

Stonehenge's enigmatic nature has made it particularly prone, and suitable, to multiple interpretations. We don't really know why it was constructed and we know very little about the people who constructed it other than they inhabited part of the same island space that is today Britain. This opens up possibilities. It could be seen as having a universality;

> being in the presence of stones that date from before the advent of nation states or any current civilization, it is possible to entertains ideas of generalized kinship and connectedness with the most fundamental aspects of humanity and of the heavens.[16]

It could be regarded as the ultimate monument of a pluralism that unites rather than divides. As no-one on this island has any real cultural link to the makers of Stonehenge, other than propinquity in space, it could be regarded as having as much relevance to recent immigrants as to those whose ancestors were immigrants in previous millennia; Stonehenge becomes a monument about being here. However, even if this was feasible and desirable it is in turn loaded with cultural assumption, such as, the assumption that all cultures root with the occupation of land. As we consider below, implicit in the notion of 'historic environment' is the idea of place. In other words heritage is perceived as being sedentary, rather than mobile and portable. It is connected to objects that are connected umbilically to a geographical location, rather than by, say the objects carried by nomads or the oral history and traditions treasured by migrants. This concept of heritage was formalized as part of the development of modern conservation practice.

Classifying Value

The practice of seeking to understand, protect and value heritage has a pre-modern ancestry as we have briefly alluded to above with the example of Stonehenge. However, heritage as we understand it today, in the West at least, is a fundamentally modern practice.

Modernity in this context refers to the broad movement rooted in the Enlightenment period of eighteenth-century Europe that was essentially secular and progressive, in the sense of seeking to break with history and tradition. It is this characteristic of modernity that is most obviously in potential conflict with the

---

16   A. Ballantyne, 'Misprisions of Stonehenge', 25, op. cit.

process of heritage, although at the same time providing the stimulus for thought and action. For example, while facilitating the industrial revolution, the Enlightenment also contributed to other processes, including the development of a modern historical consciousness and the nation-state. The new relationships with culture and religion, with nature and environment, generated new conceptions of time. History came to be interpreted as a collective social experience which recognized that different cultures and places had different natures. Historicity, the belief that each period in history has its own beliefs and values, led to a consideration of works of art and of historic buildings as unique, and so worthy of conservation as an expression of a particular culture and a reflection of national identity.[17] Furthermore, though the origins of the nation-state lie earlier, the concept crystallized in this period, aided by such traumatic events as the French Revolution. A more strongly defined nationalism, based around the territorial unit of the nation-state, demanded both a process of building identity and a common national heritage.[18] Co-existing with open, dynamic modernity with its push for newness and change is 'traditionalism', which seeks to 'harness change and re-cement identity by appealing to the authority of traditions'.[19] In the nineteenth century the painful process of primary industrialization was accompanied by an idealization of imagined pasts. David Brett links this to the nineteenth-century concern with 'national character' as part of the process of nation-building.[20] The distant past was invoked and traditions invented, such as royal pageants, as part of this construction of national heritages. In the context of Victorian England, Charles Dellheim showed how traditions were mobilized flexibly and selectively to support divergent points of view and how heritage was used to reinforce both local and national identity.[21] Thus various European countries developed legal frameworks and heritage bureaucracies to protect national heritage. At the same time there was a developing view that cultural heritage might have a universal value to mankind.

Thus established processes of heritage are usually inherently 'modern' for two reasons. First, they are a reaction to the threat caused by progressive modernity and the change (whether aesthetic or social) that this implies. The impetus towards conservation and conscious selection and retention of buildings expands with each move to demolish or alter these buildings. It is bound into a complex dialectic with change, and used to affirm the continuity and stability necessary for nationhood. Second, heritage professionals are people of the modern age. Their concepts of history and cultural value and their methods of pursuing their goals are as intrinsically modern as those of the promoters of change. For example, from an

---

17 J. Jokilehto, *A History of Architectural Conservation*, (Oxford: Butterworth Heinemann, 1999).

18 B. Graham, G.J. Ashworth, et al., *A Geography of Heritage*, (London: Arnold, 2000).

19 M. Glendinning, 'A cult of the modern age', *Context*, 68 (2000): 13.

20 D. Brett, *The Construction of Heritage*, (Cork: Cork University Press, 1996).

21 C. Dellheim, *The Face of the Past: The Preservation of the Medieval Inheritance in Victorian England*, (Cambridge: Cambridge University Press, 1982).

early period they have relied on ideas of selection and classification, eventually expressed in state-defined and controlled lists, and on principles of conservation, which though morally based, can be rationally applied by a skilled elite.

One manifestation of this is the development of classifications and typologies of heritage value. A key historical landmark in this respect is generally taken to be the typology of heritage values produced in 1903 by Alois Riegl, at the time the Austrian state-appointed 'General Conservator', who sought a more refined understanding of the motives which lay behind the process of conservation.[22] In summary, Riegl divided heritage values into two broad categories. The first was memorial values, such as age value, historical value and intended memorial value and the second present-day values, such as use value, art value, newness value, and relative art value.

Many subsequent typologies have been produced, and whilst there is some acknowledgement of the reductionist problems they can cause,[23] typologies lie perhaps even more than ever at the heart of the conservation process. Modern conservation practice seeks to determine different strands of value, which generally are labelled under a broad umbrella of cultural significance. These are then mobilized into a process of 'value-led management'.[24] It is worth briefly noting here that as typologies have evolved we can see an ever-broadening in terms of their scope, most recently embracing, for example, 'intangible values' which have traditionally lain outside Western concepts of heritage value. At the same time, whilst there has been an increasing awareness of the position that heritage values are a social construction of time and place, there has been a stubborn clinging to ideas of 'intrinsic value' and a wish to separate them from the more obviously instrumental performative roles of heritage.

Dealing With Pluralism

The development of ideas of cultural value and the contingent and variable nature of heritage have very gradually begun to permeate heritage management practice. At least some opening up of ideas of how we value heritage and whose values should inform this process has been evident. A key document in this regard, for its influence far beyond its initial national concerns if for nothing else, has been the Australian *Burra Charter*.[25]

---

22   J. Jokilehto, *A History of Architectural Conservation*, op. cit.

23   E. Avrami, et al., *Values and Heritage Conservation*, (Los Angeles: The Getty Conservation Institute, 2000).

24   D. Worthing and S. Bond, *Managing Built Heritage: The Role of Cultural Significance*, (Oxford: Blackwell, 2008).

25   Australia ICOMOS, *The Burra Charter: The Australia ICOMOS Charter for Places of Cultural Significance*, (Victoria: Australia ICOMOS, 1999), http://www.icomos.org/australia/burra.html.

The Burra Charter is a document specifically drawn up to inform the conservation of heritage in Australia. First adopted in 1979 and subsequently revised on three occasions (most recently in 1999), its popularity in the UK has principally derived from its recommendations on codifying systematic conservation process; essentially following a sequence of research and analysis before intervention and ensuring the proper documentation of interventions. A parallel manifestation of this Australian importation has been the wide-scale adoption of conservation plans as a means of understanding historic sites and informing decision-making. Within this there are some critical concepts useful for an intelligent approach to managing historic assets. Of particular value is the emphasis placed upon defining *significance*; what it is that makes a historic site important, before deciding what can be done with it. However, less commented upon is another crucial facet of *The Burra Charter* that is of interest here; the question of how conservation knowledge is defined.

The starting premise of *The Burra Charter* is that *place* is important. To understand the cultural significance of place involves an understanding of familiar elements such as the fabric and its setting and use. But this significance also stems from people's memory and association with place. Thus, judging significance is not just an architectural or archaeological appraisal of fabric, but is also reliant upon incorporating people's experience. How place is valued in conservation terms should not, therefore, be entirely through conventional expert values; although how much this occurs in actual practice, despite reference to the Charter, is questioned by Gibson discussing the Australian state of Queensland in this volume.

In the development of a specifically British discourse on conservation, English Heritage's statement, *Sustaining the Historic Environment,* was a significant document in opening perspectives on heritage issues to include non-expert opinion.[26] The document recognized the potential gap between the sort of historic environment that has been historically validated by conservation values and the types of everyday environments, which people may value and which may underpin local distinctiveness and identity. Stemming from this, *Sustaining the Historic Environment* identified the importance of recognizing non-expert values and placed an importance on securing wider public participation in conservation debates.

A recent international statement on historic environment value and conservation process has come from the Council of Europe with its *Framework Convention on the Value of Cultural Heritage for Society* (or *Faro Convention*).[27] The Convention seeks to 'square the circle' by combining universalistic notions of heritage with the idea that heritage also needs to be considered pluralistically. The universalistic claim of cultural heritage is considered at a conceptual level and linked to the

---

26   English Heritage, *Sustaining the Historic Environment: New Perspectives on the Future,* (London: English Heritage, 1997).

27   Council of Europe, *Council of Europe Framework Convention on the Value of Cultural Heritage for Society,* (Faro: Council of Europe, 2005).

concept of rights; that is, people have rights over their cultural heritage. The accompanying commentary claims this to be an innovation of the Convention.[28] It refers to the 1948 United Nations *Universal Declaration of Human Rights*, 'Recognizing that every person has a right to engage with the cultural heritage of their choice, while respecting the rights and freedoms of others...'.[29] Emphasis is placed upon pluralist democratic engagement. Thus Article 12a commits signatories to encouraging everyone to participate in:

- The process of identification, study, interpretation, protection, conservation and presentation of the cultural heritage;
- Public reflection and debate on the opportunities and challenges which the cultural heritage represents.[30]

Cultural heritage is seen as intrinsic to sustainable development, cultural diversity in the face of the threat of homogenizing globalization and a resource around which to construct dialogue, democratic debate and openness between cultures. On this last point the Convention is 'Convinced of the soundness of the principle of heritage policies and educational initiatives which treat all cultural heritages equitably and so promote dialogue among culture and religion'.[31] The Convention also acknowledges the economic role of heritage and 'without excluding the exceptional, particularly embraces the commonplace heritage of all people'.[32]

Thus, both internationally and in the UK we can see a process whereby heritage management practice has sought to democratize; to embrace more pluralistic definitions of heritage and more inclusive processes of heritage management. However, the shift to a more open and inclusive process of management is not straightforward. For instance, Emma Waterton, Laurajane Smith and Gary Campbell have identified contradictions within *The Burra Charter*.[33] Their analysis sees no resolution between its rather vague call for democratization and the continued significance placed upon traditional notions of authority and expertise. Conservation experts continue to reserve to themselves judgements over what has been historically the primary focus of conservation activity, the physical fabric.

However, if the heritage sector were to fully open up considerations of value and significance this would pose some very serious questions for heritage practice. To

---

28 Council of Europe, *Council of Europe Framework Convention on the Value of Cultural Heritage for Society: Explanatory Report*, (Faro: Council of Europe, 2005).

29 Council of Europe, *Explanatory Report*, Preamble, op. cit.

30 Council of Europe, *Framework Convention on the Value of Cultural Heritage for Society*, Article 12a, op. cit.

31 Council of Europe, *Framework Convention on the Value of Cultural Heritage for Society*, Preamble, op. cit.

32 Council of Europe, *Explanatory Report*, 4, op. cit.

33 E. Waterton, L. Smith and G. Campbell, 'The utility of discourse analysis to heritage studies: *The Burra Charter* and social inclusion', *International Journal of Heritage Studies*, 12, 4 (2006): 339–355.

take a fully post-modern approach would potentially plunge heritage management into a relativistic morass, where all values are individual rather than collective and any possibility of a shared narrative illusory. A philosophically less extreme issue for the heritage sector could be that a more pluralistic approach to heritage opens up different sorts of consensus, which are in conflict with the dominant values in the sector. For instance, in the context of the parallel activity of countryside conservation, Philip Goodwin has argued that the extension of participation in countryside conservation 'may be bringing about a retreat from the national vision of traditional conservation and a fragmentation of conservation ideas'.[34] Again, this takes heritage agencies into potentially deeply uncomfortable territory.

Waterton and colleagues suggest that whilst *The Burra Charter* encourages wider participation in defining meaning in the historic environment, it does not relinquish control from the experts over defining cultural significance, the key measure guiding management decisions. Extending this argument, how much are the efforts of the heritage sector on contemporary agendas of pluralism actually about *retaining* control in the face of political hostility? Is the sector prepared to relinquish at least a measure of control, or is the rhetoric of pluralism used as lip-service to sustain control in the face of broader political agendas? Should the various pronouncements and heritage statements discussed above be seen as part of an evolving family of documents that seek to rationalize and justify orthodox versions of the importance of heritage protection, part of Laurajane Smith's 'Authorized Heritage Discourse'?[35]

Whilst there are clear grounds for such a critique we would wish to put forward four qualifications on this. First, orthodox heritage practice *has* reformed and evolved. Whilst it is evident that, not surprisingly, the heritage sector wishes to sustain its own authority and power it is our contention that there has been some genuine efforts at a wider democratic engagement, however limited, and a degree of reflexive institutional reform, and that there is the potential for this reform to continue and evolve. We believe that constructive engagement with this reform potential requires attendance to the practical conditions for heritage's management *as well as* critical analysis of the discourses which inform it. Second, as John Pendlebury, Tim Townshend and Rose Gilroy argue in this volume in relation to Byker, the 'Authorized Heritage Discourse', in implementation at least, should not be seen exclusively as inflexible and monolithic.[36] The accepted narrative of place can develop from dialogue including powerful and articulate local voices, rather than being simply an external elite imposition. The challenge for heritage management is how to enable these dialogues and then to reflect them in practice. Third, different spheres of heritage activity have different potentials and

---

34 P. Goodwin, 'The end of consensus? The impact of participatory initiatives on conceptions of conservation and the countryside in the United Kingdom', *Environment and Planning D: Society and Space*, 17 (1999): 383.

35 L. Smith, *Uses of Heritage*, (Oxford: Routledge, 2006).

36 Pendlebury et al., this volume.

limitations. For example, the Heritage Lottery Fund has shifted priority from large capital projects of physical conservation to a greater emphasis upon community benefit and small community-led projects with no physical conservation element. Such a shift would be much harder to achieve in the domain of conservation-planning, with its extensive collections of statutorily protected assets and where the alternative to protection by cultural elites might in practice lead to local development nexuses, generally powerful in local governance, moving in, with consequent traducing or even erasure of the historic environment. Again in the conservation-planning arena, devolving power further to local communities may just result in empowering a NIMBYism that is in reality primarily concerned with other issues, such as sustaining property values. Fourth, many critiques of current heritage practice seek a more devolved, more inclusive, community-led process of heritage identification, protection and management. In addition to the problems we have already identified with this aspiration, and see Pendlebury et al and Gibson in this volume for further consideration of these issues, it also contains an implicit assumption that people *want* to be engaged in this process. The work of Nicolas Shore challenges this.[37] His interviews with Asian residents in Gloucester found many people had a high degree of respect for traditional narratives of heritage (such as the Cathedral and Docks in Gloucester) but little wish to engage. Indeed, Shore found that whatever the attitudes respondents held towards authorized heritage there was no sense of exclusionary barriers or the need for reformist inclusionary policies.

Themes and Structure of this Book

While all the chapters in this book explore typologies and constructions of value the chapters in Part I, 'Values and Heritage Stewardship', are particularly focused on this through their identification and discussion of 'crises' or, at the least, endemic failures in the stewardship of material culture. In his contribution David Lowenthal identifies a 'crisis' facing contemporary museums, which, he argues, are 'beleaguered because fast-changing views of their proper functions lumber them with multiple and ever more incompatible missions'.[38] Lowenthal argues that these challenges are not particular to museums but that the entire heritage sector 'is in a perpetual state of emergency' as it seeks to be at one and the same time responsive to the desires of the Government and at the same time retain its own internal authority and meaning.[39] For Lowenthal museums must be careful about the extent to which they attach their heritage

---

37   N. Shore, *Whose Heritage? The Construction of Cultural Built Heritage in a Pluralist, Multicultural England.* (Ph.d. Thesis, School of Architecture, Planning and Landscape, Newcastle University: 2007).
38   Lowenthal, this volume, 19.
39   Ibid., 19.

stewardship to individual or community identity as to do so risks allowing heritage to be used for 'national or tribalist aims'.[40] Rather, heritage stewardship must seek to retain the 'detached distancing' that enables its institutions to function 'as reliable vehicles of public illumination'.[41] According to Laurajane Smith this 'detached distancing' is part of a structure she has termed the 'Authorized Heritage Discourse' (AHD). Smith has discussed the dimensions of the AHD at length elsewhere;[42] in this volume she is concerned to document a particular instance of the social affect of the AHD. Smith argues that heritage is not the bearer of particular intrinsic values but is in fact a process which functions to transmit particular values, including social values, that 'society or sections of a society wish to preserve and "pass' on".[43] Smith demonstrates the social affect of this 'heritage as process' through the discussion of research undertaken with visitors to English country houses. She argues that the function of the English country house as 'heritage' is to act as a vehicle for the assertion of class hierarchy. Peter Howard is concerned with another kind of elitism, the exclusive control of heritage by experts. He argues that the construction of heritage as a concept, its designation and its contemporary management is dictated by experts – not only heritage managers but also historians, geologists and others. However, Howard does not advocate a simple replacement of 'expertise' with 'the people' but calls for experts to 'get off our seat comfortably behind Caesar, and get down into the arena and dispute with others as to the future heritage'.[44] On this reading different expert disciplines become communities whose interests can be debated on an even plane with other 'non-expert' communities.

'The historic environment' has for some time been a key term in heritage policy in England and is currently the preferred English Heritage terminology for referring to heritage, such as buildings and places.[45] In Australia too the phrase is key, for instance, the journal of Australian ICOMOS is titled *Historic Environment*. However, as Peter Howard has discussed in his contribution to this book, UNESCO and the Council of Europe have preferred the term 'cultural landscape' as in, for instance, the *European Landscape Convention*.[46] In one of the formative statements of the English 'historic environment', *Power of Place*, the starting point is the results of a MORI survey which purported to demonstrate the 'power of place'.[47] 'People care' about 'the historic environment' the document concludes in its first

40   Ibid., 29.
41   Ibid., 30.
42   L. Smith, *Uses of Heritage*, op. cit.
43   Smith, this volume, 33.
44   Howard, this volume, 61.
45   The term 'historic environment' first came to be used as a key phrase in English heritage policy in 1994 with the publication of *Planning Policy Guidance 15: Planning and the Historic Environment* (PPG15).
46   Council of Europe, *European Landscape Convention, 2000*, http://conventions. coe.int/Treaty/en/Treaties/Html/176.htm.
47   English Heritage, *Power of Place: The Future of the Historic Environment*, (London: English Heritage, 2000).

few pages.[48] From this opening position where the importance of the 'historic environment' to 'the people' is established, the document goes on to state that 'the historic environment is what generations of people have made of the places in which they lived'.[49] Thus, *Power of Place* describes a 'historic environment' in which change takes place; nevertheless, the articulation of 'historic environment' here and as it has developed since then, is of a 'historic environment' which is primarily valued for the traces of the past embodied within it. If this understanding of the past is a little more plural than perhaps twenty years ago, the nomenclature of 'historic environment' still militates against the possibility of more active and contemporary engagements with landscapes which might produce new versions of the past or by new versions of the past as represented in markers of the present.[50] The 'historic environment' therefore is constitutively limited. It is for this reason that we have titled Part II 'Cultural Landscapes'.

The chapters in Part II discuss a variety of cultural landscapes and construct their object – 'landscape' or 'environment' – in different ways. Lisanne Gibson's chapter on the cultural landscape of the Australian state of Queensland discusses the designation and management of monuments, memorials and public art. She argues that, despite heritage programmes which include a focus on cultural and social value, many communities histories and experiences are not represented by the outdoor cultural objects in Queensland which are managed as heritage. Gibson's point in this chapter is to underscore the social and political affects of heritage. John Schofield is also concerned to investigate the ways in which people engage with cultural heritage and landscape. Particularly, he is interested in exploring the material culture and landscapes which evoke 'the more personal meanings and values... than state-led mechanisms for heritage management currently allow'.[51] Like Gibson he identifies that these objects and places are often not designated as 'special' and afforded statutory protection. Schofield is not necessarily advocating statutory protection as the only or best tool for heritage's management but is interested in exploring 'why these intimate engagements matter and how they can be made more widely available to those who want to share in them'.[52] John Walton and Jason Wood in making the case for the World Heritage listing of Blackpool, 'the world's first working-class seaside resort',[53] are also interested in heritage which is more 'everyday'. Walton and Wood discuss the cultural landscape of Blackpool as 'heritage of the recent past' and consider the modalities which allow such a reading. The designation of Blackpool as a World Heritage Site is also considered in terms of its instrumental benefits; specifically

48   Ibid., 4.
49   Ibid.
50   See also Howard's discussion of the problems with the 'historic environment' terminology in this volume.
51   Schofield, this volume, 93.
52   Ibid., 94.
53   Walton and Wood, this volume, 115.

Walton and Wood argue that there would be regeneration effects, such as, a boost to the economy from increased tourism, from such a listing. Tracey Avery is also concerned with the practical dimensions of preservation in her discussion of the statutory protection of 'street art' in the laneways of the Australian city of Melbourne. In her discussion of the National Trust's (Victoria) involvement in debates about the protection of what some members of the community term 'graffiti' and others term 'art', Avery is concerned with both the historic and social values of this kind of 'street art' but also with the conceptual and practical problems for heritage management presented by such material culture.

The third section focuses on 'The Heritage of Housing'. Peter Borsay gives a long historical perspective over how something that might now seem the bedrock of a particular type of Englishness, the Georgian house, has been perceived in a variety of historically contingent ways. The status it now enjoys as a heritage icon has evolved and been actively 'made'. Both subsequent chapters in this section focus on the valorization of twentieth-century social housing as heritage. Unlike the value now attributed to the Georgian house, the protection and conservation of large-scale welfare state housing is often controversial, as it is a form of development often held in low regard by many. Inevitable debates arise over whether identifying such housing as 'special' is the imposition of an avant-garde cultural elite on residents and on the public purse. This is the primary focus of the chapter by John Pendlebury, Tim Townshend and Rose Gilroy that considers the listing of the Byker Estate in Newcastle upon Tyne and how it relates to local conceptions of place and how it might influence the future of the estate. Whilst also examining social housing, Peter Malpass delves into perhaps more fundamental issues about housing as heritage; houses have a very particular cultural construction around the idea of 'home' and the way this relates to personal identity These different ways of constructing heritage – as part of a wider architectural historical narrative or something more personal – are at the heart of Malpass's chapter, where he makes a separation between 'Heritage' and 'heritage'. He examines this distinction using the examples of the Sea Mills Estate in Bristol and the Byker Estate in Newcastle upon Tyne.

Valuing Historic Environments considers a diverse array of environments and landscapes, some of which have been officially designated 'heritage', some have not. All, however, are valued in various ways by different groups and individuals. Can 'heritage' with all its conceptual and practical limitations encompass all of these different engagements with material and indeed intangible culture? Probably not. But the constant critical engagement with the discourses which inform 'heritage', the practical considerations which constrain it and above all the cultural, social and political affects of its management (or not) is not a sign of the failure of the 'heritage' concept. On the contrary, the vigour of the debate is a sign that in addition to elitism there is also a democratic liberalism which is a central, although perhaps paradoxical, element of the heritage discourse.

Bibliography

Australia ICOMOS *The Burra Charter: The Australia ICOMOS Charter for Places of Cultural Significance.* Victoria: Australia ICOMOS, 1999. http://www.icomos.org/australia/burra.html.

Avrami, E. et al. *Values and Heritage Conservation.* Los Angeles: The Getty Conservation Institute, 2000.

Ballantyne, A. 'Misprisions of Stonehenge' in *Architecture as Experience: Radical Change in Spatial Practice* edited by D. Arnold and A. Ballantyne, 11–35, London: Routledge, 2004.

Belfiore, E. 'Art as a means of alleviating social exclusion: Does it really work? A critique of instrumental cultural policies and social impact studies in the UK', *International Journal of Cultural Policy*, 8, 1 (2002): 91–106.

Bender, B. *Stonehenge: Making Space.* Oxford: Berg, 1998.

Bennett, T. and Savage, M. 'Introduction: Cultural capital and cultural policy', *Cultural Trends* 13, 2 (2004): 7–14.

Bennett, T. and Silva, E.B., 'Cultural capital and inequality: Policy issues and contexts', *Cultural Trends*, 15, 2/3 (2006): 87–106.

Bourdieu, P. *Distinction: A Social Critique of the Judgement of Taste.* Oxford: Routledge, 1984.

Brett, D. *The Construction of Heritage.* Cork: Cork University Press, 1996.

Council of Europe *Council of Europe Framework Convention on the Value of Cultural Heritage for Society.* Faro: Council of Europe, 2005.

Council of Europe *Council of Europe Framework Convention on the Value of Cultural Heritage for Society: Explanatory Report.* Faro: Council of Europe, 2005.

Council of Europe *European Landscape Convention.* http://conventions.coe.int/Treaty/en/Treaties/Html/176.htm, 2000.

Davis, P. *Ecomuseums: A Sense of Place.* Leicester: Leicester University Press, 1999.

De La Torre, M. A*ssessing the Values of Cultural Heritage.* Los Angeles: The Getty Conservation Institute, 2002.

Dellheim, C. *The Face of the Past: The Preservation of the Medieval Inheritance in Victorian England.* Cambridge: Cambridge University Press, 1982.

English Heritage *Power of Place: The Future of the Historic Environment.* London: English Heritage, 2000.

English Heritage *Sustaining the Historic Environment: New Perspectives on the Future.* London: English Heritage, 1997.

Gibson, L. 'In defence of instrumentality', *Cultural Trends*, 17, 4 (2008): 247–257.

Gibson, L. *The Uses of Art: Constructing Australian Identities.* Brisbane: University of Queensland Press, 2001.

Gibson, L. and Besley, J. *Monumental Queensland: Signposts on a Cultural Landscape*, Brisbane: University of Queensland Press, 2004.

Glendinning, M. 'A cult of the modern age', *Context*, 68 (2000): 13–15.

Goodwin, P. 'The end of consensus? The impact of participatory initiatives on conceptions of conservation and the countryside in the United Kingdom', *Environment and Planning D: Society and Space*, 17 (1999): 383–401.

Graham, B. Ashworth, G.J. and Tunbridge, J.E. *A Geography of Heritage*. London: Arnold, 2000.

Harvey, D. *The Condition of Postmodernity*. Oxford: Blackwell, 1990.

Healey, P. *Collaborative Planning: Shaping Places in Fragmented Societies*. Basingstoke: Macmillan, 2006.

Holden, J. *Capturing Cultural Value: How Culture has Become a Tool of Government Policy*. Demos: London, 2004.

Jencks, C. *What is Post-Modernism?* Chichester: Academy Editions, 1996.

Jokilehto, J. *A History of Architectural Conservation*. Oxford: Butterworth Heinemann, 1999.

Newman, A., McLean, F. and Urquhart, G. 'Museums and the active citizen: tackling the problems of social exclusion', *Citizenship Studies*, 9, 1 (2005): 41–57.

O'Neill, M. 'Commentaries: John Holden's *Capturing Cultural Value: How Culture has Become a Tool of Government Policy*', *Cultural Trends*, 14(1), 53 (2005): 113–128.

Pendlebury, J. *Conservation and the Age of Consensus*. London: Routledge, 2009.

Pendlebury, J., et al. 'The conservation of english cultural built heritage: A force for social inclusion?', *International Journal for Heritage Studies*, 10, 1 (2004): 11–32.

Selwood, S. 'Measuring culture', *Spiked Culture*. http://www.spiked-online.com/Printable/00000006DBAF.htm, 2002, accessed 01/09/2003.

Shore, N. *Whose Heritage? The Construction of Cultural Built Heritage in a Pluralist, Multicultural England*. Ph.D. Thesis, School of Architecture, Planning and Landscape, Newcastle University, 2007.

Smith, L. *Uses of Heritage*. Oxford: Routledge, 2006.

Strange, I. and Whitney, D. 'The changing roles and purposes of heritage conservation in the UK', *Planning Practice and Research*, 18, 2–3 (2003): 219–229.

Waterton, E., Smith, L., and Campbell, G. 'The utility of discourse analysis to heritage studies: *The Burra Charter* and social inclusion', *International Journal of Heritage Studies*, 12, 4 (2006): 339–355.

West, C. and Smith, C. '"We are not a Government poodle": Museums and social inclusion under New Labour', *International Journal of Cultural Policy*, 11, 3 (2005): 275–288.

Worthing, D. and S. Bond, *Managing Built Heritage: The Role of Cultural Significance*. Oxford: Blackwell, 2008.

Worthington, A. *Stonehenge: Celebration and Subversion*. Loughborough: Alternative Albion, 2004.

PART I
Values and Heritage Stewardship

Chapter 1
Patrons, Populists, Apologists: Crises in Museum Stewardship

David Lowenthal

Museums today are more popular and at the same time more beset than ever. They are popular because increasingly visited by mass audiences, who widely regard them as trustworthy sources about everything from pterodactyls to Picassos. They are beleaguered because fast-changing views of their proper functions lumber them with multiple and ever more incompatible missions. Public approbation lends museums high visibility, while intensifying their burdens and risks.

Here I sketch the dilemmas faced by repositories of natural and cultural heritage in becoming embattled and politicized arenas of conflict. As traditional acquisition norms of purchase and pillage give way to moral and legal demands for restitution and repatriation, many museums are hard put to retain their holdings, let alone to add to them. At the same time, views about how, when and even if museums should display their wares are in flux. Past triumphalism gives way to stress on remorse and recrimination; the primacy once accorded perdurable material relics is now accorded to function and performance; elite patronage yields to dumbed-down populism; ideals of objectivity and universality are discarded for partisan engagement.

Museums are not the only heritage realm undergoing crisis. Indeed, heritage as a whole is in a perpetual state of emergency. Stewards in every realm are beset by manifold challenges: how to ensure a valued past's survival, how to fund its conservation and maintenance, how to validate its authenticity, how to secure its proper role in public memory. Perils of the moment make heritage managers more reactive than proactive; they respond when things look parlous. In so doing they mirror public awareness and concern. Nothing arouses affection for a legacy so much as the threat of its loss. Things of which most people were little aware suddenly become precious when destruction or despoliation is imminent. Few in the heritage community gave much thought to Mostar's bridge, Afghanistan's cliff-face Buddhas or Baghdad's museum until they were bombed, defaced and looted.

Yet museums especially exemplify the purposes and attendant problems of our collective heritage. More than any other institution, they epitomize what we value not as commercial commodity – although museums dote on what used to be commodities – but as locales in which to be edified, charmed, delighted, amazed, awed, challenged. Admittedly, such terms equally evoke other arenas – circuses, funfairs, theatres, sport stadiums. But museums differ from all of these in that entertainment is not their primary function. Much public pressure is nowadays exerted to make it so. Museums that fail to oblige populist demands are denounced as old-fashioned, fuddy-duddy, unresponsive, elitist.

Old Museum Missions and Stereotypes

Things were ever thus. Every era is lumbered with once-trendy museums that then became white elephants – obsolete embarrassments, politically incorrect sitting ducks for iconoclasts. They are in unending trouble. 'Museums anywhere, everywhere, are a problem', declared heritage guru Freeman Tilden twenty years ago. In the light of 'widespread public resistance to museums of any kind', he went on, 'it has even been suggested that the name museum be changed to something else'.[1]

Part of this opprobrium stemmed from the pre-modern European collecting urge that begat *Wunderkammerer*, those omnium gatherums whose bric-a-brac influence still persisted. Tilden had toured countless 'historical-society museums that left me dazed and dizzy, ... a letter from Napoleon to Josephine reposing beside a stuffed albino squirrel ... a fireman's hat just above a first edition of James Fenimore Cooper, and both of these in front of a Revolutionary musket ... Each article displayed was ... possibly a treasure, but the whole set-up was inchoate'.[2]

Yet cabinets of curiosities, the private collections of rich and powerful men, were by then the exception. They had been largely transformed into or replaced by nineteenth-century public museums, serving as sanctuaries for relics of past times and other cultures, at once exotic, instructive and, as vestiges of outworn folk doomed to extinction, pathetic. The great Western museums came into existence as showcases of imperial booty, taken by force or theft or derisory purchase from colonized and subjugated peoples the world over. As adjuncts of national and imperial chauvinism, the public acquisition of global heritage justified looting and spoliation in the name of civilized progress. Termed 'the finest legacy of the Enlightenment',[3] the modern universal museum was a temple of knowledge and beauty intended to inspire patriotic reverence and wholesome reflection.

---

1 Freeman Tilden, 'That Elderly Schoolma'am: Nature', in Freeman Tilden, *The National Parks*, rev. ed., ed. Paul Schullery (New York: Alfred A. Knopf, 1951), reprinted in Freeman Tilden, *Interpreting Our Heritage*, 4th ed., ed. Bruce Craig (Chapel Hill: University of North Carolina Press, 2008), 166–73 at 172.

2 Tilden, 'That Elderly Schoolma'am' (see note 1), 173. Similar complaints had been levelled against London's Royal Society Repository more than two centuries earlier: 'not only in no sort of order or tidiness but covered with dust, filth and coal-smoke' (Zacharias Conrad von Uffenbach (1710)), quoted in Giles Waterfield, 'Anticipating the Enlightenment: Museums and Galleries in Britain before the British Museum', in R. G. W. Anderson et al., *Enlightening the British: Knowledge, Discovery and the Museum in the Eighteenth Century* (London: British Museum Press, 2003), 5–10 at 6.

3 Alan Riding, 'The Industry of Art Goes Global', *New York Times*, 28 March 2007.

Moral pedagogy was the canonical museum role. The mission of their collections was to teach a wider public to emulate their betters. The likenesses of admired exemplars in the National Portrait Gallery would inspire 'mental exertion, noble actions and good conduct', was Prime Minister Lord Palmerston's 1856 dictum inscribed above the portal.[4] Viewing the artefacts of other cultures would make British Museum visitors more thoughtful, more benign, more truly British.

Like the temples of worship they were often built to resemble, these secular cathedrals aimed to inculcate civic ideals among the untutored masses, but only to hoi polloi able and ready to shed uncouth ways. 'The first function of a Museum is to give example of perfect order and perfect elegance to the disorderly and rude populace', wrote that arbiter of taste John Ruskin in 1880. But the populace should pay for the privilege of self-improvement. There should be 'small entrance fees, that the rooms not be encumbered by the idle, or disgraced by the disreputable. You must not make your Museum a refuge against rain or ennui, nor let into perfectly well-furnished and even … palatial rooms the utterly squalid and ill-bred portion of the people'.[5] Museums epitomized decorum and civility – the antithesis of Bateman's Boy, jailed for life for breathing on the glass.[6] Museum treasures were sacred icons, not to be touched, let alone doubted or derided.

Small wonder that to their critics museums were akin to mausoleums. In 1856 the British Museum stifled Nathaniel Hawthorne.

> Crushed to see so much, I wandered from hall to hall with a heavy and weary heart, wishing that the Elgin Marbles and the frieze of the Parthenon were all burnt into lime … and that the mummies had all turned to dust … that all the material relics of so many successive ages had disappeared with the generations that produced them.

Misguided veneration of antiquity, in his view, blinded viewers from seeing 'for rubbish what is really rubbish; and under this head might be reckoned almost everything … at the British Museum'.[7]

Half a century later, 'museum' became Italian Futurists' prime term of abuse. Everything wrong with modern Italy came from its fetish of museums, their useless,

---

4   Quoted in Robert Lumley, *The Museum Time-Machine: Putting Cultures on Display* (London: Routledge, 1988), 224.

5   John Ruskin, 'A museum or picture gallery: Its function and formation' (*Art Journal*, 1880), in *The Complete Works of John Ruskin*, ed. E.T. Cook and Alexander Wedderburn (London: George Allen, 1908), 34: 247–62 at 249–50. Although open to the public from 1759, the British Museum long required advance applications from prospective visitors. The first genuinely open museum was the Louvre, free to all after 1793. But working-class and 'respectable' audiences were everywhere segregated.

6   H.M. Bateman, *The Boy Who Breathed on the Glass in the British Museum* (*Punch*, October 1916; London: Methuen, 1921).

7   Nathaniel Hawthorne, *The English Notebooks*, 29 Sept. 1855 (New York: Modern Language Association of America, 1941), 242–3.

filthy, retrograde holdings the major impediment to modern progress. Italy was a 'country of the dead', its people dozing over the glory of their ancestors, Rome and Venice mired in mouldy relics, Florence a cemetery kept up for tourists besotted with the antiquarian rubbish. In seeking to rid Italy of 'its smelly gangrene of professors, archaeologists, ciceroni and antiquarians', Filippo Tommaso Marinetti delighted in the 'growing nausea for the antique, for the worm-eaten and moss-grown. ... For too long has Italy been a dealer in second-hand clothes. We mean to free her from the numberless museums that cover her like so many graveyards'.[8]

Anti-museum stereotypes continue to proliferate. The museum as mortuary, as site of death and entombment, starred in Dan Brown's best-selling *Da Vinci Code*.[9] Yet the word most commonly linked with museums is 'boring'. Modern Londoners are said to see the British Museum as 'dusty, irrelevant and dull, ... the place of boring school trips'.[10] It is also stupefying in its immensity. 'Teenagers looking for ancient artefacts have to face an expedition almost as fraught as Indiana Jones's adventures in the Temple of Doom'.[11] In New York's Metropolitan Museum of Art, a fleet-footed guide promises to deliver the top dozen masterpieces in under an hour. The champion of 'the six-minute Louvre' sprints past the *Venus de Milo*, the *Winged Victory of Samothrace*, the *Mona Lisa*, exulting that 'there isn't a museum in the world that can keep me inside for very long'.[12]

New Strengths and Obligations

Today's museums signally refute these denigratory stereotypes. Museums are more popular, more numerous, more visited, more extolled, more trusted than ever. They are prime stimulants of domestic pride and foreign tourism; they are founts of group identity; they become envoys of diplomacy. And they are seen as the most trustworthy source of public instruction; the public at large has greater faith in museums than in any other supplier of knowledge. In a 1994–1995 survey, 20,000 European 15-year-olds ranked museums and sites most reliable for understanding the past – above documents and printed sources, much higher than TV documentaries.[13] Across the Atlantic the story is the same: 1,500 American

---

8 'Founding and manifesto of Futurism' (1909); 'Birth of a Futurist aesthetic' (1911–1915); 'Founding and manifesto', in *Marinetti: Selected Writings*, ed. R.W. Flint (London: Secker & Warburg, 1972), 43, 81, 42.

9 A.O. Scott, 'Was it kind of quiet at the museum? Dead, actually', *New York Times*, 18 May 2006, B1: 6.

10 Jayne Dowle, 'The British Museum should recognise that the hold of history operates more subtlety than the gee-whizzery of science', *The Times*, 1 April 1999.

11 Ibid..

12 Art Buchwald, 'The six-minute Louvre', *International Herald Tribune*, 10–11 March 1990: 6.

13 Signe Barschdorff, 'Is history teaching up to date? Introduction', in Joke van der Leeuw–Roord, *The State of History Education in Europe* (Hamburg: Körber-Stiftung, 1998),

interviewees considered museums the most honest and unbiased source of information about the past, more trustworthy even than accounts by grandparents and by eyewitnesses to historic events.[14]

What makes museums so trusted? They provide unparalleled intimacy and involvement with the past, in apparently unmediated immediacy. In museums we go at our own pace, with our own chosen companions; we are not required to look here rather than there, or constrained to view things in any set sequence; we are not told what to think but left to make up our own minds. Unlike students immured in a classroom or imbibing oldsters' tales of yore, museum-goers feel they make of the past what they themselves decide, based on evidence that is seemingly uncontrived and objective.[15]

Intrinsic materiality also lends verisimilitude to museum displays. Visitors trust what is there because they see it with their own eyes; seeing is believing, things don't lie, as is customarily said. Imaginations are carried back to when the bones were living creatures and the artefacts were fashioned, without the distortions patent in movies and television.[16] Moreover, museum exhibits are felt to reflect a consensus of many views, not just one, as with a particular schoolteacher or an authored textbook. The collaborative museum chorus of manifold voices, all the more authoritative by dint of their anonymity, lends displays added credibility.[17]

Public approbation in great measure reflects the extraordinary transformations museums have undergone in recent decades. Victorian credos are now stood on their heads. Populism dethrones patronage. Populist taste more and more decides

---

77–92 at 84; Angela Kindervater and Bodo von Borries, 'Historical motivation and historical-political socialization', in *Youth and History: A Comparative European Survey on Historical Consciousness and Political Attitudes among Adolescents*, ed. Magne Angvik and Bodo von Borries, 2 vols. (Hamburg: Körber-Stiftung, 1997), A 62–105 at 87; B 45. Scores, on a 5-point scale, were museums 4.15, books and documents 3.93, television 3.64, films 2.81, novels 2.72.

14   Roy Rosenzweig and David Thelen, *The Presence of the Past: Popular Uses of History in American Life* (New York: Columbia University Press, 1998), 21. On a ten-point scale, museums rated 8.4 against 8.0 for grandparents, 7.8 for eyewitnesses, 6.6 for high school teachers, 5.0 for movies and television.

15   Ibid., 195.

16   Ibid., 32, 105–6.

17   Ibid., 108. Faith in museums is often unwarranted. The public believes museums partly because they are often compelled to display the version of history it prefers. Nor does expert consensus guarantee objectivity or impartiality. And admirers of museums' 'unmediated' past scenarios do not know, or conveniently forget, that what they see is distorted by selection, placement, signage, cleaning, restoration, juxtaposition and context. Faith in material veracity is also delusive. Tangibility lends credence to physical relics. Here we are, they seem to say; you can see us, even touch us; why doubt the reality of your senses? Yet tangible museum exhibits, however compelling, are incomplete and deficient. From material remains alone we can merely speculate about past minds, hearts and memories; only recorded words, whether feigned or sincere, reveal conscious intent, forethought and hindsight.

what is to be acquired and shapes its display – motorcycles at the Guggenheim, hip-hop ephemera at the Brooklyn Museum of Art, 1950s' fashion at the Victoria and Albert. Vox populi makes the take at the gate the prime criterion of museum survival, confirming Andy Warhol's 'canny prediction that all museums will become department stores, and vice versa'.[18] Ancient relics give way to retro nostalgia, contemplation to performance, display cases to interactive engagement; several art museums encourage visitors to add their own labels to paintings and sculpture.[19]

At the conceptual artist Rudolf Stingel's retrospective in Chicago and New York, viewers were invited to have a go at gallery walls with pens, credit cards, fingernails, or whatever; the resultant 'populist, manic, talking-in-tongues wallpaper' was said to make 'the Metropolitan Museum's Temple of Dendur, incised by a couple of centuries of tourist comments, look positively virginal'.[20] Media hypes blockbuster exhibits; what is permanent is musty and often politically incorrect. Display features more and more objects from the present and the very recent past, a time near enough to the present for visitors to identify with. A 2006 British Museum show on the modern Middle East foregrounded 'works that directly address the issues that exercise us all', wrote the museum's director: 'political and religious turmoil, violence, displacement, exile, the struggle for liberties of all kinds'.[21]

Allied with interaction and recency is the rising devaluation of artefact durability, even of the primacy of artefactual holdings. Once the sine qua non of museums and the very stuff of heritage, material objects increasingly take a back seat to fetishes of performance and interaction, to virtual substitutes and to concern with intangible heritage. Process is privileged over product, function over preservation. Removing an object from active involvement in ongoing social exchange deprives it of meaningful life and impairs the society from which it came. At the same time, the cult of evanescence means that many new artistic creations come with provisos to let them decay and expire rather than be conserved. It is symptomatic of these trends that the recent debate entitled 'Should we junk collections?' took place at the Wallace Collection, one of London's premier permanent museum showcases.[22]

Eschewing the collectors' mindset (though not their money), museums become exemplars of self-denying altruism. If few private collectors yet 'worry about the

18   Holland Cotter, 'Leaving room for the troublemakers', *New York Times*, 28 March 2007, H1: 35. The ensuing irony is that the Pittsburgh museum dedicated to Warhol seems like an exhausting department store (Brooks Adams, 'Industrial-Strength Warhol', *Art in America*, 82: 9 [Sept. 1994], 70–7, 129–31 at 129. On museum competition with shopping malls, movie theatres, even grocery stores, see Carol Vogel, 'The art of listening', *New York Times*. 12 March 2008: H 1, 15.

19   Pamela Licalzi O'Connell, 'One Picture, 1,000 Tags', *New York Times*, 28 March 2007: H39.

20   Roberta Smith, 'Making their mark', *New York Times*, 13 Oct. 2007: A19.

21   Neil MacGregor in *The Guardian* (London), quoted in Alan Riding, 'Museum with mission: Global enlightenment', *International Herald Tribune*, 2 June, 2006.

22   The Wallace Collection, London, 16 May 2005.

future, when public opinion might turn against the cultural havoc generated by the rape of archaeological sites', museums are increasingly called upon, even required, to let things go.[23] Repatriating a Ghost Dance shirt to the Lakota tribe in North America, Glasgow Museums termed the loss of the shirt outweighed by 'bringing healing to a sad people'.[24] In negotiating with tribal and developing-world claimants, museums become front-line emissaries of collective national consciences. Canadian museum personnel engage in social healing by returning potlatch hoards, salvaging totem poles, marketing Inuit soapstone sculpture and curating creation myths. Trumpeting its 'historic openness to the world', the British Museum vaunts 'an innovative and subtle form of cultural diplomacy' aimed at spreading 'awareness and acceptance of cultural diversity'.[25]

Many new museums are avowedly didactic, partisan, chauvinist. They are impelled to become centres for social change, 'places where a cultural identity is hammered out, refined and reshaped, [and] community centers, where a group gathers to celebrate its past, commemorate its tragedies and convey its achievements to others'.[26] Like other stewards and communicators of the past, museums align themselves more and more with tragedy not victory, with the horrors of victims rather than the triumphs of the powerful. Launched in 2002, the International Coalition of Historic Site Museums of Conscience embraces museums of slavery, the Irish Famine, the Holocaust, the Gulag, tenement slums, sites of incarceration (prisons, workhouses), and scenes of slaughter.[27] A 'Troubles' museum in Northern Ireland has been proposed to mark Catholic–Protestant reconciliation.[28] (Trouble is trendy: a Museum of Broken Relationships in Berlin displays mementoes cast off by forsaken lovers – wedding dresses, old underwear, a coffee machine, a prosthetic limb that 'lasted longer than their love, as it was made of better material' – to purge memories of discarded partners.[29]) To achieve empathetic involvement, visitors take on the personae of specific slaves at the Charles Wright Museum of African American History in Detroit, of doomed Jews at the US Holocaust Memorial Museum, Washington, DC, gaining emotional engagement through visceral contact with replicas of whips, chains and doomed victims' clothes and shoes.

23   Souren Melikian, 'The endgame in Chinese art', *International Herald Tribune*, 3–4 April 1999.

24   Mark O'Neill quoted in Josie Appleton, *Museums for 'The People'?* (London: Institute of Ideas, 2001), 18.

25   Riding, 'Museum with mission' (see note 20).

26   Appleton, *Museums* (see note 22), 21–2; Edward Rothstein, 'Anecdotal Evidence of Homesick Mankind', *New York Times*, 20 July 2006: B1.

27   Ruth J. Abram, 'Kitchen conversations: Democracy in action at the Lower East Side Tenement Museum', *The Public Historian* 29:1 (2007), 59–76 at 65.

28   Richard Morrison, 'When abstinence makes the arts grow fonder (and better)', *The Times* (London), 26 Sept. 2006: T2: 7.

29   Bojan Pancevski, 'Your Broken CDs and Torn Suits are in the Museum', *The Times* (London), 20 Oct. 2007: T2, 2.

In hosting contrition, the museum becomes a holy sanctuary, a sacred site of cathartic healing, a locus for 'an ethically charged experience, a psychologically fraught encounter, a stage for disruptive, possibly dangerous, ideas'.[30] At Birmingham's Museum and Art Gallery, a textile training centre for 'isolated' Asian women not only helps 'improve their skills and self-confidence, but it also provides a safe space for mental health issues to be confronted and discussed'.[31] At the same time, museums are used to spearhead urban and regional economic development. Thus Bilbao's new Guggenheim Museum served as a 160 million dollar shot-in-the-arm in tourist revenues.[32] It is abundantly clear that 'museums no longer have the luxury of being passive warehouses of the world's accumulated knowledge and material record', as Inter-American museum spokesmen said. They 'must foster sustainability in the local community'.[33]

New Dilemmas and Difficulties

These changes have done much to make museums more popular. But they have not, on the whole, made them more affluent, confident or trustworthy. The physical and financial difficulties associated with looking after and safeguarding fabric and collections are more formidable than ever. Museum income seldom matches escalating costs. Risks of theft, iconoclasm, accident, natural disaster and human mishap require huge expenditures on safety precautions and incur crippling insurance charges. Insurance has become virtually impossible for travelling exhibitions subject to ever greater hazards, including possible seizure and incarceration from suits by claimants contesting ownership. Increasingly, museums pare back funding for ever more expensive conservation and acquisition. Only a fraction of the cost of coping with crowds of often non-paying visitors is met by sales in museum gift shops and cafes. And more guards, warders and cleaners mean fewer curators and less research.[34]

Policies that mandate decay and restitution pose painful dilemmas for curators schooled to husband treasured works for as long as possible, and for museum trustees who consider such works capital assets whose loss is insufferable. Hence decay is apt to be proscribed even when dissolution was the creator's intent. Works

---

30 Cotter, 'Leaving room for the troublemakers' (see note 17), H1. See Victoria Newhouse, *Towards a New Museum* (New York: Monacelli Press, 1998).

31 Group for Large Local Authority Museums, *Museums and Social Inclusion* (2000), quoted in Appleton, *Museums* (see note 23), 17.

32 T.R. Reid, 'Revival Built on a Glittering Museum', *International Herald Tribune*, 1 April 1999.

33 American Association of Museums, *Museums and Sustainable Communities* (Washington and Costa Rica, 1998), 1.

34 Ann V. Gunn and R.G.W. Prescott, 'Lifting the veil: Research and scholarship in United Kingdom museums and galleries' (1999), in Appleton, *Museums* (see note 23), 20.

whose *raison-d'être* is transience are conserved against artists' express wishes, lest their loss violate public trust and jeopardize museum funding. Artists who want their creations to decompose and perish are at times overruled by curators who regard accession as an eternal act, by conservators who aspire to permanence and by trustees intolerant of capital loss. Museums remain dedicated to the fiction that works of art are fixed and immortal.[35]

Restitution is no less problematic. Demands for repatriation are increasingly heeded, new acquisitions shunned. Post-imperial morality frowns on clinging to old or acquiring new exotic treasures. The traditional term 'keeper' now suggests an anal-retentive hoarder of other people's stuff, much of it stored out of public sight. Today, national export codes and international ethical diktats restrict what museums may buy or be given. What they got earlier is besieged by repatriation claims. Classical art and antiquities flow back to Italy and Greece and Egypt from the Metropolitan Museum of Art and the Getty, if not yet from the British Museum and the Louvre. Bones and artefacts restituted to tribal indigenes in America and Australia deplete museums the world over. UNESCO-backed demands from ex-colonial and other claimants sap curatorial energies, threaten promised donor legacies and undermine public support for museum stewardship.

And objects that remain in museum hands are less and less freely displayed. Universal access is abrogated in favour of ethnic, tribal and religious limitations; secrets must be kept from outsiders. 'Australian male totems are barred from female eyes' at the Hancock Museum in Newcastle upon Tyne; at the British Museum only priests are allowed to view the wooden tablets Ethiopian Christians believe represent the original Ark of the Covenant. The UK Museums Association *Code of Ethics* urges 'restricting access' to ceremonial and religious items 'where unrestricted access may cause offence or distress to actual or cultural descendants'.[36] How, indeed, do you 'build a museum to display cultures which have a deep ambivalence about the notion of being displayed?'[37]

Avowed partisanship tarnishes museums' reputation for being trustworthy. Consider the fate of a recent Petrie and Croydon Museums/Burrell Collection touring exhibition:

> 'Digging for Dreams' was ostensibly an exhibition of Egyptian artefacts. But it also had as its goal the promotion of social and cultural inclusion in economically deprived areas. You might wonder what ancient Egypt had to do with such a worthy cause. The curators however managed to twist a connection together: to challenge the Eurocentric portrayal of Ancient Egypt and replace it with an

35   Ann Temkin, 'Strange fruit', in *Mortality Immortality? The Legacy of 20th-Century Art*, ed. Miguel Angel Corzo (Los Angeles: The Getty Institute, 1999), 45–50.

36   Tiffany Jenkins, 'The censoring of our museums: The back half', *New Statesman*, 11 July 2005 <http://www.newstatesman.com/Arts/200507110035>.

37   Alison Landsberg, *Prosthetic Memory: The Transformation of American Remembrance in the Age of Mass Culture* (New York: Columbia University Press, 2004), 82, 129–34.

Afrocentric one. The idea was that black visitors would feel better about their own history, and therefore better about themselves.[38]

Museums can never hope to tell the whole truth or the sole truth of any past. Nor can they ever escape civic demands that require them to curate useful fictions and selective forgettings. They can only try to resist such pressures as best they may, or succumb to iconoclastic despair.

In Siegfried Lenz's novel *The Heritage*, invading German and Russian forces recurrently restage exhibits in a Masurian museum to demonstrate folk origins and links with conquering Teutons or Slavs. Masurian survivors eventually flee to Schleswig, rebuilding their museum as a homeland memento. But German and then Polish chauvinism continues to deform what is displayed. Finally the curator torches the whole museum, so that 'the collected witnesses to our past … could never again be exploited for this cause or that'.[39] Even more than in Lenz's tale, the actual past is now openly exploited in museum display. Congressional pressures reshaped the Smithsonian's ill-famed *Enola Gay* exhibition: from enlightening visitors about much-debated consequences of the Hiroshima bombing, it became a patriotic show designed to please American veterans.[40] Washington's new Museum of the American Indian makes Native Americans feel good by simplifying history, glorifying racial essentialism and replacing controversy with moral righteousness. 'It's an Indian thing', a visitor is told; 'you wouldn't understand'.[41] Museum codes of ethics increasingly exhort shaping and restricting access to displays in accordance with the beliefs and feelings of cultural descendants, privileging the idea that truth and authority are vested in blood and belief. It was once taken for granted that it was museums' job to help us understand; now it is to instil faith.

Honouring Long-term Responsibilities

Museums uniquely mediate past, present and future. Ensuring that today's viewpoints do not swamp tomorrow's is a thankless task in museums buffeted by 'ephemeral media pressure and political clamour', as a British Museum director put it.[42] It is all too easy nowadays to pillory stewardship as hoarding. But memorabilia

---

38   Ian Walker, 'The curator's tale', in Appleton, *Museums* (see note 23), 31.

39   Siegfried Lenz, *The Heritage* (London: Secker & Warburg, 1981), 458.

40   Roger D. Launius, 'American memory, culture wars, and the challenge of presenting science and technology in a national museum', *The Public Historian* 29: 1 (2007), 13–30 at 19; Martin Harwit, *An Exhibit Denied: Lobbying the History of Enola Gay* (New York: Copernicus, 1996).

41   Stephen Conn, 'Heritage vs. history at the National Museum of the American Indian', *The Public Historian* 28: 2 (2006), 69–73 at 72.

42   David M. Wilson, ed., *The Collections of the British Museum* (London: British Museum Press, 1989), 13.

of no immediate moment may be our best defence against public amnesia. To serve posterity museums must have a measure of custodial autonomy. To abnegate all aloofness, to respond only to immediate needs, defeats our ultimate interests. It maroons us in a shallow present devoid of temporal insight.

Museums came into existence expressly to collect and protect relics from the past for display to present and future viewers. Populist immediacy – heeding vox populi – seems democratic, but it disfranchises the great majority: posterity yet to come. The more museums bend to present demands, the less they can heed our heirs. When funds are cut, the future gets short shrift.[43]

Museums today must heed new masters, new merits, new remits. But though driven by new mandates, they remain committed to, indeed often shackled by, the old ones. They are expected to undertake modern roles, as exemplars of selfless scruples, of fleeting popular taste, of trendy relevance, and at the same time to go on sustaining traditional functions: gathering in the best of everything, mirroring national goals, safeguarding stored sources of memory and inspiration, truth and beauty. The old verities – valued canon, accepted hierarchy, immortal treasures, civic edification – no longer figure in museum mission statements. Yet they remain expectations deeply felt, not only by specialists but by the general public. If museums did as Hawthorne proposed, all hell would break loose. To sell off, give away or junk holdings would unleash storms of protest. The public wants it both ways, not to be bored by paragons yet to know they are safely and permanently in place. These abiding aims collide with insistent current demands for public involvement and social amelioration. As the former director of the National Portrait Gallery and the National Gallery puts it:

> Those who use museums and visit them don't want them used instrumentally. They want them to be open-ended, under-programmed places for individual experience. They want them to be places of history and culture and public education. … They don't mind if museums are about subjects that they don't fully understand. That is why they come to them.[44]

Conclusion

Our age is awash in a surfeit of heritage. Museums are alike beneficiaries of our obsession with heritage, and among its complicit promoters. Museums' new-found sensitivity to empathy leaves them at the mercy of those who would bend them to national or tribalist aims or, still worse, enlist them in the generalized politics of memory, which sacralizes the emotional salience of remembrance to the detriment of

43   David Lowenthal, 'From patronage to populism', *Museums Journal*, 92:3 (March 1992), 24–27.

44   Charles Saumarez Smith, 'A challenge to the new orthodoxy', in Appleton, *Museums* (see note 23), 39.

historical understanding. Empathetic concerns have their place. But they should not be allowed to overshadow the detached distancing that enables museums uniquely to serve, and to be widely seen, as reliable vehicles of public illumination.

Bibliography

Abram, Ruth J. 'Kitchen conversations: Democracy in action at the Lower East Side Tenement Museum', *The Public Historian* 29: 2 (2007), 59–76.
Adams, Brooks. 'Industrial-strength Warhol', *Art in America* 82: 9 (Sept. 1994), 70–7, 129–31.
American Association of Museums. *Museums and Sustainable Communities*. Washington and Costa Rica, 1998.
Appleton, Josie. *Museums for 'The People'?* London: Institute of Ideas, 2001.
Barschdorff, Signe. 'Is history teaching up to date? Introduction', in *The State of History Education in Europe* edited by Joke van der Leeuw–Roord, 77–92. Hamburg: Körber-Stiftung, 1998.
Bateman, H.M. *The Boy Who Breathed on the Glass in the British Museum. Punch,* October 1916; London: Methuen, 1921.
Buchwald, Art. 'The six-minute Louvre', *International Herald Tribune.* 10–11 March 1990, 6.
Conn, Stephen. 'Heritage vs. History at the National Museum of the American Indian', *The Public Historian.* 28: 2 (2006): 69–73.
Cotter, Holland. 'Leaving Room for the Troublemakers', *New York Times,* 28 March 2007, H1: 35.
Dowle, Jayne. 'The British Museum should recognise that the hold of history operates more subtlety than the gee-whizzery of science', *The Times,* 1 April 1999.
Harwit, Martin. *An Exhibit Denied: Lobbying the History of* Enola Gay. New York: Copernicus, 1996.
Hawthorne, Nathaniel. *The English Notebooks.* New York: Modern Language Association of America, 1941.
Jenkins, Tiffany. 'The censoring of our museums: The back half', *New Statesman,* 11 July 2005 <http://www.newstatesman.com/Arts/200507110035>.
Kindervater, Angela and von Borries, Bodo. 'Historical Motivation and Historical-Political Socialization', in *Youth and History: A Comparative European Survey on Historical Consciousness and Political Attitudes among Adolescents,* edited by Magne Angvik and Bodo von Borries, 2 vols, A 62–105. Hamburg: Körber-Stiftung, 1997.
Landsberg, Alison. *Prosthetic Memory: The Transformation of American Remembrance in the Age of Mass Culture.* New York: Columbia University Press, 2004.
Launius, Roger D. 'American memory, culture wars, and the challenge of presenting science and technology in a national museum', *The Public Historian* 29: 1 (2007): 13–30.

Lenz, Siegfried, *The Heritage*. London: Secker & Warburg, 1981.

Lowenthal, David. 'From patronage to populism', *Museums Journal*, 92: 3 (March 1992): 24–27.

Lumley, Robert. *The Museum Time-Machine: Putting Cultures on Display*. London: Routledge, 1988.

Marinetti, Filippo Tommaso, *Marinetti: Selected Writings* edited by R.W. Flint. London: Secker & Warburg, 1972.

Melikian, Souren. 'The endgame in Chinese art', *International Herald Tribune*, 3–4 April 1999.

Morrison, Richard. 'When abstinence makes the arts grow fonder (and better)', *The Times* (London), 26 Sept. 2006: T2: 7.

Newhouse, Victoria. *Towards a New Museum*. New York: Monacelli Press, 1998.

O'Connell, Pamela Licalzi. 'One picture, 1,000 tags', *New York Times*, 28 March 2007: H39.

Pancevski, Bojan. 'Your Broken CDs and Torn Suits are in the Museum', *The Times* (London), 20 Oct. 2007: T2, 2.

Reid, T.R. 'Revival built on a glittering Museum', *International Herald Tribune*, 1 April 1999.

Riding, Alan. 'Museum with mission: Global enlightenment', *International Herald Tribune*, 2 June, 2006.

Riding, Alan. 'The Industry of Art Goes Global', *New York Times*, 28 March 2007.

Rosenzweig, Roy and Thelen, David. *The Presence of the Past: Popular Uses of History in American Life*. New York: Columbia University Press, 1998.

Rothstein, Edward. 'Anecdotal evidence of homesick mankind', *New York Times*, 20 July 2006: B1.

Ruskin, John. 'A museum or picture gallery: Its function and formation' (*Art Journal*, 1880), in *The Complete Works of John Ruskin* edited by E.T. Cook and Alexander Wedderburn, 34: 247–62. London: George Allen, 1908.

Scott, A.O. 'Was It Kind of Quiet at the Museum? Dead, Actually', *New York Times*, 18 May 2006, B1: 6.

Smith, Roberta. 'Making their mark', *New York Times*, 13 Oct. 2007: A19.

Temkin, Ann. 'Strange fruit', in *Mortality Immortality? The Legacy of 20th-Century Art* edited by Miguel Angel Corzo, 45–50. Los Angeles: The Getty Institute, 1999.

Tilden, Freeman. 'That Elderly Schoolma'am: Nature' in Freeman Tilden, *The National Parks*, rev. ed., edited by Paul Schullery. New York: Alfred A. Knopf, 1951, reprinted in Freeman Tilden, *Interpreting Our Heritage*, 4th ed., edited by Bruce Craig, 166–73. Chapel Hill: University of North Carolina Press, 2008.

Vogel, Carol. 'The art of listening', *New York Times*, 12 March 2008: H 1, 15.

Waterfield, Giles. 'Anticipating the enlightenment: Museums and galleries in Britain before the British Museum' in *Enlightening the British: Knowledge, Discovery and the Museum in the Eighteenth Century* edited by R.G.W. Anderson et al., 5–10. London: British Museum Press, 2003.

Wilson, David M. ed., *The Collections of the British Museum*. London: British Museum Press, 1989.

Chapter 2

Deference and Humility:
The Social Values of the Country House

Laurajane Smith

The idea of 'value' is central to both the idea of heritage and in framing conservation policies and practices. Heritage is defined as something of value, and particular heritage sites are placed on historic registers or other 'lists' of important objects and places because of the values they are deemed to represent and 'have'. Heritage is also conserved, managed and preserved according to the values – the cultural significance – that a heritage place or object is assessed as having, and, ideally, the cultural significance or value of heritage should determine how it is used, managed or conserved.[1] Social value, one of the themes of this volume, has often been defined in the heritage literature as a particular set of values within the wider pantheon of heritage values.[2] However, this chapter argues that heritage does not 'have' value, but rather heritage is a cultural process that is about re/creating, negotiating and transmitting certain values – among them social values – that society or sections of a society wish to preserve and 'pass' on. This is an argument about the nature of heritage that I have developed elsewhere,[3] however, this chapter explores the idea that heritage has a social affect through the way the social value of heritage is defined and experienced. The management of heritage sites and their interpretation and use as recreational places, and the various assumptions that are made in these processes about the nature of heritage and the values it has, work to help regulate contemporary social values and the social identities that are

---

1   For discussion of the role of 'cultural significance' in heritage management and conservation, see: M. Pearson and S. Sullivan, *Looking After Heritage Places* (Melbourne: Melbourne University Press, 1995); J.S. Kerr, *The Conservation Plan* (4th edn, Sydney: NSW National Trust, 1996); K. Clark, editor, *Conservation Plans in Action: Proceedings of the Oxford Conference* (London: English Heritage, 1999); Australia ICOMOS, *Charter for the Conservation of Places of Cultural Significance (Burra Charter)*, 1999; English Heritage, *Conservation Principles for the Sustainable Management of the Historic Environment* (London: English Heritage, 2006); D. Worthing, and S. Bond, *Managing Built Heritage: The Role of Cultural Significance* (Oxford: Blackwell, 2008).

2   See the following for debates about 'social' value: C. Johnston, *What is Social Value?* (Canberra: Australian Government Publishing Service, 1992); Pearson and Sullivan *ibid*; D. Byrne et al. *Social Significance: A Discussion Paper* (Sydney: Department of Environment and Conservation NSW, 2001); and English Heritage ibid.

3   L. Smith, *Uses of Heritage*, (London: Routledge, 2006).

informed by or are underpinned by them. In short, heritage is a cultural process or performance, concerned ultimately not with the management of things, but with the management and regulation of social value and cultural meanings.

This chapter examines the role heritage plays in not only symbolizing certain social values, but also the way it is used in reinforcing, renegotiating and disseminating those values in English society generally. The country house, although often defined as iconic of England's national heritage, reinforces and continually recreates a particular suite of social values that speak to particular sections of English society.[4] The country house, or house museum, has also been extensively criticized for its conservative, if not reactionary, imagery and symbolism.[5] These tensions make the country house a useful example to demonstrate the affect that the social values of heritage can have. Although this chapter concentrates on the country house, the process of recreation of the social values both of the heritage site, and those of importance to a particular social order, are simply what all heritage does – whether it be a country house, a terraced house, industrial site, archaeological monument or whatever – although the particular values that are recreated and negotiated will obviously differ.

I argue there is a dominant way of talking, thinking about or valuing heritage, which I refer to as the 'Authorized Heritage Discourse'. This discourse structures both the ways in which the country houses are interpreted to and by visitors, and how the social values of class deference both underlie, and are reproduced, in the interpretations of the houses that visitors engage in. The messages that visitors take away from the house, or that are reinforced by their visit, are transmitted into wider English society, and this has important consequences for class identity in England and the social values that inform those identities.

What is Heritage?

This chapter commences with the observation that the value of any item or place of heritage is not innate or inherent, but is assigned to heritage places by people. This

---

4   The status of national icon of the country house/stately home is discussed and asserted in much of the literature on the history and meaning of the country house/stately home, see for instance: S. Andreae, and M. Binney, *Tomorrow's Ruins?*(London: SAVE Britain's Heritage, 1978); S. Andreae et al. *Silent Mansions: More Country Houses at Risk* (London: SAVE Britain's Heritage, 1981); M. Binney, *Our Vanishing Heritage* (London: Arlington Books, 1984); A. Tinniswood, *A History of Country House Visiting: Five Centuries of Tourism and Taste* (Oxford: Blackwell, 1989); P. Mandler, *The Fall and Rise of the Stately Home* (New Haven: Yale University Press, 1997); J. Cornforth, *The Country Houses of England 1948–1998* (London: Constable, 1998).

5   See for instance: P. Wright, *On Living in an Old Country* (London: Verso, 1985); R. Hewison, *The Heritage Industry: Britain in a Climate of Decline* (London: Meltuen London Ltd., 1987); K. Walsh, *The Representation of the Past: Museums and Heritage in the Post-Modern World* (London: Routledge, 1992).

may seem like an obvious point, but it is a point often obscured in the way heritage is traditionally defined and talked about. In the traditional or dominant discourses of heritage in England – the Authorized Heritage Discourse (or AHD) – heritage is defined as being all that is 'good', grand, monumental and, primarily, of national significance. Above all the AHD has naturalized the assumption that the values of heritage – the values that make it good, grand or nationally significant – are inherent to the heritage object and place. However, what makes a place, building or artefact 'heritage' is the symbolic role it is *given* both within and through the processes of remembering and commemoration. This role is to symbolize or even stand in for the values that a certain society or section of society hold as important and worthy of retention if not veneration.

Though the Authorized Heritage Discourse constructs heritage as an object or place to be managed and conserved, this chapter also incorporates a definition of heritage oppositional to the one propagated by the English AHD.[6] Heritage is therefore not seen as a thing – it is not a country house, or other place, monument or artefact – but rather what happens at and with those places, monuments or artefacts. Heritage is a process or a performance, or what Bella Dicks may call an act of communication,[7] undertaken at certain sites or places – places that are themselves identified as important because of the performances that occur at them. It is a process that involves commemoration, memorializing and most importantly collective and individual remembering. Above all, it is a process in which social and cultural meanings and values are identified, considered, recreated, rejected or otherwise negotiated. The identification of heritage places, their management and conservation and their interpretation to visitors is itself a process and performance of meaning making in which certain heritage sites, and the social and cultural values and meanings they are considered to represent, are continually reinforced and recreated. This is done at a national level, as organizations such as English Heritage and the National Trust define what sites are to be raised to the symbolic role of heritage. The sites they choose speak to, and in turn legitimize and preserve, the values that are collectively seen by these organizations as important in underpinning national as well as certain regional and community identities. This process is also done at more intimate levels, as visitors to heritage sites themselves participate in the recreation of social values and meanings through the performance of visiting, and the collective and individual rememberings and commemorations that are undertaken during the visit.

However, this process does not of course occur in a vacuum and it is itself regulated or governed by a particular discourse or conceptual construction of

---

6   For further discussion of the Authorized Heritage Discourse and a wide-ranging discussion of the definition and idea of heritage used in this chapter, see Smith 2006 op. cit., also E. Waterton, L. Smith, and G. Campbell, 'The utility of discourse analysis to heritage studies: *The Burra Charter* and social inclusion', *International Journal of Heritage Studies* 12 (2006): 339–355.

7   B. Dicks, *Heritage, Place and Community* (Cardiff: University of Wales Press, 2000).

heritage. The AHD structures the way meanings and values are recreated, while also validating or legitimizing certain values and meanings and de-legitimizing others. This is not to say that the AHD is all powerful, and forms an exclusive vision of heritage and the values heritage places may hold. There are certainly many examples of community heritage groups who work outside or in opposition to the AHD.[8] However, the AHD and the ways it naturalizes certain heritage values and conceptualizations of heritage has become embedded and is continually rehearsed within public policy in England.[9] The pervasiveness of this discourse also, however, works to obfuscate the process of meaning making. It does so by reinforcing the idea that social value and cultural meaning are inherent in a heritage place, and that, ultimately, it is the heritage place and its fabric that are all important and the focus of our attention rather than the way that place is used. This discourse developed in Western Europe in the nineteenth century, and one of its foundational principles or ethics derives from Ruskin's and Morris's idea of 'conserve as found'. Encapsulated in this conservation ethic is not only the idea of the inherent value of material heritage, but also the important role that they play in representing the social and cultural values of their times. Such a degree of importance is placed on the materiality of Western culture and its technological and material achievements, and so much is simply 'understood' by the aesthetic worth of certain objects that the material has often come to stand in for the social or cultural values it symbolizes. In other words, the monument tends to be conflated with the cultural and social values that are used to interpret it and give it meaning. Subsequently within the AHD heritage *is* the monument, archaeological site or other material thing or place, rather than cultural values or meanings. This has often meant that, as Robert Hewison has asserted,[10] heritage is about cultural stasis and backward glances. The preservation and conservation of material heritage becomes the preservation and conservation of certain desired values and cultural meanings – the meanings that material things stand in for – and Western heritage thus rightly may be criticized for is inherent conservatism. Further, much of the values and meanings underpinning the English AHD draws on master narratives of nation and class on one hand, and technical

---

8   See for instance Smith 2006 op. cit. and also: J. Littler and R. Naidoo, editors *The Politics of Heritage: The Legacies of 'Race'* (London: Routledge, 2005); E. Waterton, 'Whose sense of place? Reconciling archaeological perspectives with community values: Cultural landscapes in England', *International Journal of Heritage Studies* 11 (2005), 309–326; M. Mellor and C. Stephenson, 'The Durham Miner's Gala and the Spirit of Community', *Community Development Journal* 40 (2005): 343–351; L. Dodds et al. 'Industrial identity in a post-industrial age: Resilience in former mining communities', *Northern Economic Review*, 37 (2006): 77–89.

9   See Smith 2006 op. cit and E. Waterton, 'Rhetoric and "Reality": Politics, Policy and the Discourses of Heritage in England', (Ph.D., University of York, 2007), for detailed documentation of how and why the AHD has become embedded in English public policy.

10   Hewison 1987 op. cit.

expertise and aesthetic judgments on the other. One of the monuments that most symbolizes the AHD within England and the meanings and values under pinning it, is the country house. It is a flagship of heritage tourism and is so ubiquitously present in the English landscape and heritage discourse that it has become banal – in the sense used by Michael Billig.[11]

Billig argues that it is the every day objects, the banal, rather than the grand or monumental that most successfully symbolize and thus convey a sense of identity. Despite the monumentality of the country house, it has become a banal representation of not only national identity, but of certain social identities structured around class. They may be identified as banal as their ubiquitous presence in the landscape renders them commonplace. Brown signs pointing to heritage attractions are a significant and commonplace feature of the English landscape, the country house, of which at least 350 are currently open to public visits,[12] are themselves a brown signed ubiquitous presence. The country house may also be understood as banal in the way they are presented to visitors. The minimalist interpretation that occurs at English country houses is a feature of the genre.[13] Typically, genealogical and biographical information will be given on 'the family' or 'families' that have owned the house or paid for the house to be built, remodelled and decorated. Traditionally, however, little detailed explanation of the social, economic and political historical context of the house and its owners are given. In general, relatively little information is given on estate workers, servants, and/or slaves, although in more recent years certain individual houses have worked to increase the amount of historical information about workers and slaves. The general lack of signage/interpretive boards in country houses is, however, a common policy decision as signage may detract from the ambience and presumed authenticity of the house. Nor is it often seen as necessary as those with the cultural capital to appreciate the aesthetic qualities of the house are deemed sufficiently educated or well read to understand its history and aesthetic meanings and values. Country houses are generally marketed to attract a certain clientele, pricing policies and entrance fees are set to encourage a largely middle class visitor base that have been identified as being able to afford typical entrance fees.[14] The standard country house experience relies on the ability and readiness

---

11   M. Billig, *Banal Nationalism* (London: Sage, 1995).

12   This is a minimal estimated number reported by the Historic Houses Association, which represents privately owned stately homes only, and does not include those managed or owned by English Heritage, National Trust, city councils and other organizations. Historic Houses Association Home Page, 2006 <http://www.hha.org.uk/metadot/index.pl> (accessed 22 April 2008).

13   The minimalist interpretation of the English country house is also a feature of the house museum internationally, see for instance: P. West, 'Uncovering and interpreting women's history at historic house museums', in *Restoring Women's History through Historic Preservation*, edited by G. Dubrow and J. Goodman (Baltimore: The John Hopkins University Press, 2003).

14   S. Markwell, et al. 'The changing market for heritage tourism: A case study of visitors to historic houses in England', *International Journal of Heritage Studies*, 3 (1997): 95–108.

of the visitor to understand and intuit the historical and social values of the house. Indeed the visitor is encouraged to amble around the grounds and the house, view the artwork and artefact and curio collections and soak in the atmosphere while 'reading' the inherent meaning of the house.

So how do visitors 'read' the messages, values and meanings of the country house? As I have argued elsewhere,[15] they tend to do so within the confines of the AHD. However, the AHD does not just simply define how the house should be viewed and valued, in guiding and framing the country house performance it also guides and defines the social consequence of that performance. The next section illustrates how people construct the value of the country house and the social meanings they draw from this. The country house marketed and defined as national heritage is certainly perceived by many visitors as symbolic of their national heritage, and a strong sense of nationalism and English patriotism is expressed by visitors and discussed below. However, community and personal identity are also constructed by visitors at these houses, although the sense of community – or sense of place – constructed here centres on membership of a certain sense or understanding of what it means to be 'English middle-class'. Important in this construction is the idea of the innate social value of the country house which works to legitimize and naturalize the social values underpinning both national and class/community identities.

The Country House Affect

To determine the social consequence of heritage, and the country house in particular, interviews with visitors to country houses were undertaken; initially in 2004, with two smaller follow up surveys undertaken in 2007. In 2004, 454 visitors were interviewed at six country house sites: Nostell Priory in Yorkshire and Waddesdon Manor in Buckinghamshire, both National Trust properties, Audley End in Essex (see Figure 2.1), Brodsworth Hall in Yorkshire and Belsay Hall in Northumbria (managed by English Heritage), and Harewood House, Yorkshire. This latter house is managed by an independent trust and has been identified as one of the ten 'treasure houses of England'. In 2007, 276 visitors were again interviewed at Broadsworth Hall and Harewood House, and at Temple Newsom (Leeds City Council) and Burton Constable (independent trust) both in Yorkshire.[16] Interviews were designed to examine the memory and identity work that visitors to these places undertook during their visits, and the types of social messages and values that they constructed

15   Smith 2006 op. cit.

16   Acknowledgment and thanks is given to the managers and other personnel at each of the houses named above who allowed me and the research teams to interview their visitors both in 2004 and 2007. My thanks to Jean Hunter who invited me to undertake the 2007 Work and Play fieldwork. The 2004 work was funded by the British Academy; the 2007 work on the 1807 exhibition was funded by an AHRC Knowledge Transfer Fellowship.

Figure 2.1   An example of an English country house – Audley End, Essex, English Heritage property
Source: © Emma Waterton

and carried away from these places. The interview schedule for both 2004 and 2007 varied slightly, but both were designed to explore the memory and identity work visitors were undertaking during their visit to the house. The interview schedule consisted of both demographic questions and open-ended questions that delved into such things as how visitors defined heritage, what messages, impressions or cultural or social meanings they obtained from coming to the house they were visiting, the feelings engendered and the experiences they sought. Interviews were conducted one to one and were recorded through note taking (2004) or were recorded and transcribed (2007). For each of the open-ended questions themes were identified, codes defined, and the codes entered into a Statistical Package for the Social Sciences (SPSS) database to produce descriptive statistics.

The following discussion relates specifically to the 2004 study; the broad results of the 2007 are raised at the end of this section. The demographics of the 2004 interview population were broadly similar to the visitor demographics recorded for country houses generally. That is, the survey sample tended to conform to the socio-economic indicators traditionally associated with the middle class, and almost half had been educated to university level. For instance, 75 per cent occupied a range of managerial and professional or similar occupations, while 47 per cent had been educated to university level.[17]

---

17   Smith 2006 op. cit, 131–2.

We started the interviews by asking people what the word heritage meant to them. Many people (48 per cent) gave us various versions of the standard or authorized definition of heritage – that is these responses tended to reflect the traditional ideas of the grandiose, material, monumental and aesthetically pleasing. However, for 33 per cent of respondents in 2004, heritage was defined not so much as a thing, but as the *act* of preservation. Thus, heritage was not just a thing, or the country house, but it was also the process by which you saved or preserved it. Further, many saw themselves as engaging in heritage, engaging in the preservation of a lifestyle, building or social history, and ultimately their own social and/or national identity. This sense of engagement was defined by the act of their visit and their monetary contributions to and/or or membership of the National Trust or English Heritage – of those interviewed in 2004, 48 per cent belonged to these or similar heritage amenity societies. Many respondents saw their membership of these organizations, or the fact that they paid money to get into the house, as making a contribution to the preservation of the site. Moreover, the visit itself was perceived as a way of keeping the values that the houses stood for going:

> Everyone should see and understand it to be part of heritage that is to see how valuable it is to continue the tradition of conserving. (CH442, female, 30–39, teacher, 2004)

> The links within the family are important – it's something in their blood that needs to be preserved, opening these places to the paying public preserves them – paying our entry fee means doing our duty to the past. (CH123, female, over 60, telephonist, 2004)

There are two interesting implications raised by the above for understanding the way social value is both created and maintained by the heritage performance. First, is the sense of active engagement people had with the houses and heritage generally. Within the AHD that frames heritage/museum interpretation practices there is a dominant perception that visitors to museums and heritage sites are passive receptors of the interpreter's or curator's message.[18] This, for instance, is also an assumption of much of the criticism of heritage commercialization and sanitation led by the historian Robert Hewison.[19] Visitors, however, were not passive receptors of the country house message, but were actively engaged in the performativity of their visit. This sense of engagement is present in how visitors read the meanings and messages of the houses – these visitors were not submissive in the reading of the interpreter's message – but actively engaged in creating their own meanings and social values that helped affirm their sense of identity and

---

18   R. Mason, 'Museums, galleries and heritage: Sites of meaning-making and communication', in *Heritage, Museums and Galleries: An Introductory Reader*, edited by G. Corsane (London: Routledge, 2005).

19   Hewison 1987 op. cit.

place in the world. Secondly, not only was heritage seen as an active process of preserving things and the values they represented – these definitions also often became interlinked with peoples' perception of their own identity, and what the country house visit meant to the reinforcement of their own sense of self and identity. The identity being constructed by many of the visitors we interviewed was that of the English middle class, or at least a certain segment or vision of it. Being at, being able – or as it was often expressed, having the 'privilege' to visit – a country house and to demonstrate one's ability to understand its aesthetic social messages and values was a commemoration or demonstration of one's English middle-class identity.

This identity was constructed through a sense of the authenticity of the country house, and the comfort people drew from its stability and enduring presence. One of the major concerns of the heritage industry critique led by Patrick Wright and Robert Hewison was with the apparent lack of historical and cultural authenticity of heritage attractions, and country houses have come under quite sustained criticism in this regard.[20] What emerges from the interviews is that a very different idea of authenticity is used by respondents than that which usually concerns heritage and museum professionals. For country house visitors, the performance that is the country house visit was seen as 'authentic', as it provoked feelings and emotions that were seen as 'real' or genuine and that helped people feel 'comfortable' about their social experiences, social position and values and their sense of community. Gaynor Bagnall has also demonstrated in relation to museums that authenticity, and thus visitor engagement, often centres on the sense of the authenticity of the feelings generated by the visit.[21] The heritage performance was made 'real' or authentic not simply through the authority placed on historical 'accuracy', but on the degree to which the country house performance 'spoke' to people's emotional commitment to the social values the house as 'heritage' represented. It provided a sense of place that respondents felt comfortable enough with or sufficiently engaged with to explore, express or reflect upon their individual or collective memories, social values and identities. What is important to note is that the sense of place that emerges from responses was often uncritical, and embraced the sense of place as constructed by the way the house was managed and presented – this is not to say that it was passive, but rather tended to buy into the authorized social values of the AHD.

One of the dominant motifs to emerge from visitor surveys in both 2004 and 2007 was the sense of 'comfort' that people drew or constructed from the country house performance. The well kept and carefully controlled aesthetic environments of the country house and associated landscapes are a particular characteristic of the house museum. Indeed both Emerick and Waterton have noted that the controlled, neat and clean presentation of heritage in England is part

---

20   Wright 1985 op. cit, Hewison ibid.

21   G. Bagnall, 'Performance and performativity at heritage sites', *Museum and Society* 1 (2003): 87–103.

of the symbolism that denotes 'here be heritage'.[22] The controlled cleanliness and neatness of the presentation of heritage places is historically embedded in heritage management and interpretation polices and practices in England[23] and helps declare and underpin the value with which heritage is imbued. Subsequently heritage places generally – and country houses in particular – are physically comfortable places to visit. However, visitors also reported that they are socially 'comfortable' places as well. The country house visit was often defined as involving a feeling of comfort and security, and for some this sense of comfort was defined by the fact that they expected to encounter 'like people' at the country house. That is people socially like themselves who were undertaking similar activities as themselves at these sites. As one person put it the house was 'well looked after – people enjoy it. Very middle class here' (CH367, male, 18–29, musician, 2004). While another noted that the 'structure – formal gardens' made her 'feel special, and is an antidote to dressing down so prevalent nowadays' (CH11, female, 40–59, civil engineer, 2004).

This sense of comfort went hand in hand with the identification of the country house performance – or visit – as middle-class activity:

> As well as being in touch with heritage – it's very important part of leisure time – very middle class thing to do. ... Particularly important to middle class – gives pleasure. But that's alright different places appeal to different people. (CH369, female, over 60, academic, 2004)

> To the vast majority it doesn't mean a thing – people would rather go shopping. It seems to be a middle class thing [visiting country houses] due to education and how you are brought up to reflect, it reflects the direction of your education. (CH409, male, over 60, 2004)

The performance of the visit was an expression of membership of the English middle class. This expression was intimately tied up with the emotional responses and experiences of simply being at a country house. When visitors were asked about the emotions produced by the visit a range of emotional themes were expressed, although the persistent themes in people's responses were feelings of comfort, humility, nostalgia and nationalism.[24] Of these, by far the most frequent was comfort. This sense of comfort was inevitably tied to a sense of social security – of not only knowing your social place in the world, but also knowing the value of identifying and recognising that place:

22   K. Emerick, 'From Frozen Monuments to Fluid Landscapes. The Conservation and Preservation of Ancient Monuments From 1882 to the Present', (Ph.D. thesis, University of York, 2003); E. Waterton, 'Sights of sites: Picturing heritage, power and exclusion', *Journal of Heritage Tourism* 4(1) (2009): 37–56.
23   Emerick, ibid, 113.
24   Smith op. cit. 144.

Gives a sense of comfort – the history and stability and continuance of it. I like old houses – interested in how people lived, not what people lived – the servants and owners, the interdependence of their lives ... [it is also] comforting to know that it is still being preserved thanks to English Heritage and the National Trust. (CH286, female, over 60, 2004)

I like the house – it is warm and welcoming. I feel comfortable and at home here. (CH269, female, 30–39, computer systems operator, 2004)

Proud and comfortable. (CH363, female, 40–59, teacher, 2004)

Contented – wouldn't change my lot for this. (CH329, male, over 60, company director, and who identified that their mother had been 'in service', 2004)

Whenever I am in these places, I feel a sense of calm and belonging. (CH365, female, over 60, housewife, husband electrical engineer, 2004)

Comfortable about visiting even though it was built on slavery, but nonetheless it's part of the country's history. (CH122, male, 30–39, night shift team leader, 2004)

This sense of comfort was being used by visitors to actively help them affirm the social values associated with the country house and to construct and assert a sense of identity. A relaxing or 'nice day out' in safe and comfortable surroundings, and the sense of comfort engendered by that, is affirming a sense of middle class identity. This is occurring even when, in the case of CH122, the historical associations of place may be 'uncomfortable' – as witnessed by his acknowledgement of the close association of slavery with the house he was visiting. Here the emotional sense of security and comfort is influencing the processes of remembering and forgetting, and the social meanings being constructed by this process. In the country house performance the more traumatic and discomforting aspects of history are being rendered either invisible and thus easily forgotten, or acceptable as in the case of CH122. The country house, it needs to be pointed out, has a brutal history – the transatlantic slave trade underpins the wealth of a significant number of houses, and all such houses owe their construction and maintenance to the semi-feudal exploitation of servants and estate workers, and the wealth accrued through the owners' prominent roles in British imperialism and colonialism. Nonetheless, the country house is often portrayed as iconic of what it means to be 'British', or more particularly 'English', and they are nationally characterized as 'treasure houses' and cherished for the continuity of social values that they are seen to represent.[25] The feelings of comfort and safety expressed by many respondents were often also associated with feelings of awe and humility at the grand landscapes, vistas, and architectural wonders. In turn these feelings were expressly linked with quite reactionary social messages and meanings tied to

---

25   Mandler 1997 op. cit.; Cornforth 1998 op. cit.

aristocratic deference and the idea that things were 'better back then'. The reactionary deference of the country house is perhaps best summed up by this visitor:

> It's part of modern England as our history is part of being England. We would be still in the slums without places like this, *but it gives us something to tip our hats to* – it allows us to belong to both sides of history. (CH128, emphasis added, male, 40–59, teacher, 2004)

This sense of deference works to both identify to visitors who they are not, and who they are, and their social and cultural place in the world. They are not part of the elite – after all, they are *visitors* who have been granted the privilege to visit. They are, however, middle class as it is this class who has been given the privilege, who has the means to pay for it, and has demonstrated their worthiness for the privilege through their membership of English Heritage or the National Trust. This membership and the visit to the house is a demonstration of the possession of a set of heritage values that were seen as particular to the middle class. As one respondent noted, 'I like the National Trust it makes you feel part of a club' (CH238, male, 40–59, fire fighter, 2004). This is because preserving the heritage of the country house and caring about its preservation was defined as a middle class concern and activity: 'I don't know that Tom, Dick or Harry cares [about the house]. National Trust members are largely middle class [they care]' (CH29, male, over 60, teacher, 2004). Their place in the world is defined through their class identity – many visitors are undertaking a performance of quite literally remembering their place – their social place in the British cultural hierarchy.

Class identity was often also linked to national identity and visitors used the country house performance to remember and reinforce their national identity and the place of England in the Western cultural world. As one respondent put it,

> Stately homes are important to being English. (CH79, female, 30–39, general manager, 2004)

Although national identity was an important aspect of the visit – this was never as frequently discussed as class identity. What is important here, however, is the way the social value of the country house was used to define national identity. Frequently visitors would nominate other European countries and the United States in particular as not having what the English 'have'. This was often expressed in two ways, either other countries had not saved or preserved, and thus treasured, their heritage as the English had, and/or they lacked the social and historical continuity as represented by the country house.

> A continuum – a continuing history: America does not have a heritage. (CH29, male, over 60, teacher, 2004)

> English Architecture – something the USA does not have. Something that belongs to us. (CH21, female, over 60, Retired RAF, 2004)

Unlike USA – we are keeping history – it's not Disneyland – it's British – set out of buildings which will always be there unlike today's buildings. (CH135, male, 18–29, radio producer, 2004)

The social values of the country house are the significant issue here. Even though other European countries and the United States have their own elite house museums they are perceived as lacking the social value and continuity of the English house. England did not suffer the French and other European revolutions and this sense of continuity is often emphasized in the country house literature.[26] Nor do the English trivialize their importance – this is not Disneyland after all. The performance of preservation and conservation is, as discussed above, an important aspect of English national identity – the English conserve their past.[27] Or, more specifically, the *English* middle class conserve the English past. What of course, they are conserving, however, is not simply 'a house', but the social values that underpin middle-class deference, social position and place, and, by inference, the social and political position of the elites.

Although social deference was a dominant theme with country house visitors, there was, nonetheless a reassuring undercurrent of dissonance most often expressed when respondents were asked about the interpretation at sites. In response to questioning about this, 21 per cent of visitors requested more information about servants and/or estate workers.[28] In addition, a small percentage of visitors did identify discomfort with the social messages they were engaging with at the country houses – although this was a small minority.[29]

The strength of the social values and meanings of the country house visit was re-examined in 2007 when 185 interviews were undertaken at three county houses that were mounting exhibitions entitled 'Work and Play'. These exhibitions not only explored the leisured activities of the elites, but also aimed to reveal how that lifestyle was underwritten by the labouring activities of servants and estate workers.[30] In short, these exhibitions attempted to make these workers more visible within the country house performance. While some individuals engaged with the exhibitions and considered that they did add positively to the house experience, the majority, however, found the exhibitions either irrelevant or vaguely interesting (if they even noticed the exhibition – 33 per cent of those surveyed nominated

---

26   D. Pearce, *Conservation Today* (London: Routledge, 1989).

27   The idea that the act of managing and conserving heritage is part of English identity has been discussed by the following: Hewison 1987 op.cit.; B. Goody, 'New Britain, new heritage: The consumption of a heritage culture', *International Journal of Heritage Studies* 4 (1998): 197–205; P. Schwyzer, 'The scouring of the white horse: Archaeology identity and heritage', *Representations* 65 (1999): 42–62.

28   Smith op. cit. 158

29   Ibid., 155–158

30   See the Yorkshire Country House Partnership website for details about these exhibitions: <http://www.ychp.org.uk/maids/exhibitions.php> (accessed 22 April 2008).

that they did not notice the exhibition even though this was physically difficult to avoid). In effect, the country house performance as recorded in 2004 was not altered for the majority of people by the addition of active interpretations of the lives of the servants and estate workers.

A further 91 people were also interviewed at Harwood House in 2007 in response to an exhibition that was being mounted to mark the bicentenary of the 1807 *Act* abolishing Britain's involvement in the transatlantic slave trade.[31] While people's response to the work and play exhibition was often mild disinterest or simple lack of engagement, the response to the 1807 exhibition at Harewood House was often actively very distancing, with some visitors actually angry at being confronted with 'uncomfortable' issues as they toured the country house. A theme that emerged in response to 1807 at Harewood was the degree to which some visitors felt the bicentenary had been 'over done':

> I don't think it should be celebrated really, I don't think it deserves that much attention. (HHA11a, male, 25–34, civil servant, 2007)

> Its something in the past really, we organized it and ended it and that's it, I don't think we should dwell too long on it. (HHD11, female, over 65, retired tutor, 2007)

> I think it probably is an important part of our history, but totally overplayed at the moment ... but the other thing is apologizing, the Africans should be apologizing to themselves, if they didn't catch the slaves we wouldn't have shipped them. (HHD17, male, 45–54, insurance broker, 2007)

> It's been a little bit overplayed. Yes, it's a significance of a remembrance of what went on, I just think it's over the top. (HHD18, female, 45–54, administrator, 2007)

The 1807 bicentenary was being marked nationally throughout 2007 and thus was harder to ignore as an intervention into the country house performance than the Work and Play exhibitions. Moreover, the 1807 commemorations required some remembering of slavery – a far more 'uncomfortable' activity than the remembering of servants and estate workers. People *actively* did not like their comfort and the comfortable values of the country house being threatened by the memories of 1807. The invisibility of servants and estate workers has a long tradition not only in county house interpretation, but also in historical actuality – a 'good' servant was often defined by their invisibility. Thus, the performativity of the country house could genteelly ignore or render invisible the attempted interventions of the

---

31   This work was done as part of a larger project examining responses to the 1807 bicentenary, for details see the 1807 Commemorated project website: <http://www.history. ac.uk/1807commemorated/>.

Work and Play displays. However, the 1807 exhibition in the context of national bicentennial debates and commemorations presented a far more destabilizing intrusion to the country house performance and thus was actively resisted by the majority of the 91 visitors surveyed in 2007. For some the House's exhibition on 1807 was seen as irrelevant to the country house experience:

Interviewer: Did you find the exhibition [on 1807] interesting?

No, not really it is irrelevant; we came to see the house, the history of Princess Mary, the royal family, so it's a very kind of separate issue to visit here, so we kind of passed it by. (HHE 13, male, 35–44, architect, 2007)

Its irrelevance lies not in the fact that Harwood House does not have a long and significant connection to slavery,[32] but that it was irrelevant to the social values and sense of national and social identity that was being created, performed and remembered by the country house performance. The performance and performers can 'pass by' the opportunity to remember certain histories, because the social values that that remembering would require are simply not part of the country house performance.

Conclusion

Heritage does cultural and social work in any society – it is a focal point around which people attempt to remember and define who they are and are not. The social values represented by the country house, which are continually recreated in the performance of its management and public interpretation, represent quite reactionary social values and cultural meaning. Themes of comfort, humility, deference and most interestingly the sense of associating with 'like people' emerged forcefully from the interviews in 2004 and again in 2007. People at these sites were not only constructing a sense of their national and personal identities at these houses, but also reinforcing and legitimizing, both at a personal and collective level, the social values that endorse these identities and make them socially meaningful and real. The comfortable affirmation and rehearsing of these social values through the visit is an emotionally powerful phenomenon as evidenced by the treatment the attempts to alter the country house performance received. Attempts at inserting opportunities to remember and consider the lives of servants, estate works and slaves were quietly or actively rejected because they destabilize, if not call into question, the social values being rehearsed by the performance of both conserving and of visiting the

---

32   For this history see J. Walvin, *Britain's Slave Empire* (Stroud: Tempus Publishing, 2000); A. Tyrrell and J. Walvin, 'Whose history is it? Memorialising Britain's involvement in slavery', in *Contested Sites: Commemoration, Memorial and Popular Politics in Nineteenth Century Britain*, edited by P. A. Pickering and A. Tyrrell (Aldershot: Ashgate, 2004).

country house as heritage. The visitor as both audience and performer within the overall country house performance participates in a collective experience that, as Abercrombie and Longhurst[33] point out in their work on audiences, are able to 'diffuse out' into everyday life to inform and define middle class membership, identity and social value. The consequences of this are not simply a cosy reaffirmation of social identity, but rather a disabling reaffirmation of a social order that reinforces social and cultural exclusion and marginalization. The country house performance that 'passes by' opportunities to remember the history and legacies of servants and slaves recreates a set of social values that need not consider the realities and experiences of multiculturalism, racism and class discrimination.

Bibliography

Abercrombie, Nicholas, and Brian Longhurst. *Audiences: A Sociological Theory of Performance and Imagination*. London: Sage, 1999.

Andreae, Sophie and Marcus Binney. *Tomorrow's Ruins?* London: SAVE Britain's Heritage, 1978.

Andreae, Sophie, Marcus Binney, and Catherine Griffiths. *Silent Mansions: More Country Houses at Risk*, London: SAVE Britain's Heritage, 1981.

Australia ICOMOS. *Charter for the Conservation of Places of Cultural Significance (Burra Charter)*, 1999.

Bagnall, Gaynor. 'Performance and performativity at heritage sites', *Museum and Society* 1 (2003): 87–103.

Billig, Michael. *Banal Nationalism*, London: Sage, 1995.

Binney, Michael. *Our Vanishing Heritage*, London: Arlington Books, 1984.

Byrne, Denis, Helen Brayshaw, and Tracey Ireland. *Social Significance: A Discussion Paper*, Sydney: Department of Environment and Conservation NSW, 2001.

Clark, Kate, editor. *Conservation Plans in Action: Proceedings of the Oxford Conference*, London: English Heritage, 1999.

Cornforth, John. *The Country Houses of England 1948–1998*, London: Constable, 1998.

Dicks, Bella. *Heritage, Place and Community*, Cardiff: University of Wales Press, 2000.

Dodds, Lynn, Mary Mellor, and Carol Stephenson. 'Industrial identity in a post industrial age: Resilience in former mining communities'. *Northern Economic Review* 37 (2006): 77–89.

Emerick, Keith. 'From Frozen Monuments to Fluid Landscapes. The Conservation and Preservation of Ancient Monuments From 1882 to the Present'. Ph.D. thesis, University of York, 2003.

---

33   Nicholas Abercrombie and Brian Longhurst, *Audiences: A Sociological Theory of Performance and Imagination* (London: Sage, 1998).

English Heritage. *Conservation Principles for the Sustainable Management of the Historic Environment*, London: English Heritage, 2006.

Goody, Brian. 'New Britain, new heritage: The consumption of a heritage culture', *International Journal of Heritage Studies* 4 (1998): 197–205.

Hewison, Robert. *The Heritage Industry: Britain in a Climate of Decline*. London: Methuen London Ltd, 1987.

Johnston, Chris. *What is Social Value?*, Canberra: Australian Government Publishing Service, 1992.

Kerr, James S. *The Conservation Plan*, 4th edn, Sydney: NSW National Trust, 1996.

Littler, Jo and Roshi Naidoo, editors. *The Politics of Heritage: The Legacies of 'Race'*, London: Routledge, 2005.

Mandler Peter. *The Fall and Rise of the Stately Home*, New Haven: Yale University Press, 1997.

Markwell, Susan, Marion Bennett, and Neil Ravenscroft. 'The changing market for heritage tourism: A case study of visitors to historic houses in England', *International Journal of Heritage Studies,* 3 (1997): 95–108.

Mason, Rhiannon. 'Museums, galleries and heritage: Sites of meaning-making and Communication' in *Heritage, Museums and Galleries: An Introductory Reader*, edited by Gerard Corsane, 200–214, London: Routledge, 2005.

Mellor, Mary and Carol Stephenson. 'The Durham Miner's Gala and the Spirit of Community', *Community Development Journal* 40 (2005): 343–351.

Pearce, David. *Conservation Today*, London: Routledge, 1989.

Pearson, Michael and Sharon Sullivan. *Looking After Heritage Places*, Melbourne: Melbourne University Press, 1995.

Schwyzer, Philip. 'The scouring of the white horse: Archaeology identity and Heritage', *Representations* 65 (1999): 42–62.

Smith, Laurajane. *Uses of Heritage*, London: Routledge, 2006.

Tinniswood, Adrian. *A History of Country House Visiting: Five Centuries of Tourism and Taste*, Oxford: Blackwell, 1989.

Tyrrell, Alex and James Walvin. 'Whose history is it? Memorialising Britain's involvement in slavery', in *Contested Sites: Commemoration, Memorial and Popular Politics in Nineteenth Century Britain*, edited by Paul A. Pickering and Alex. Tyrrell, 147–169. Aldershot: Ashgate, 2004.

Walsh, Kevin. *The Representation of the Past: Museums and Heritage in the Post Modern World*, London: Routledge, 1992.

Walvin, James. *Britain's Slave Empire*, Stroud: Tempus Publishing, 2000.

Waterton, Emma. 'Whose sense of place? Reconciling archaeological perspectives with community values: Cultural landscapes in England'. *International Journal of Heritage Studies* 11 (2005), 309–326.

Waterton, Emma. 'Rhetoric and "Reality": Politics, Policy and the Discourses of Heritage in England', Ph.D. thesis, University of York, 2007.

Waterton, Emma. 'Sights of sites: Picturing heritage, power and exclusion', *Journal of Heritage Tourism* 4(1) (2009): 37–56.

Waterton, Emma, Laurajane Smith, and Gary Campbell. 'The utility of discourse analysis to heritage studies: *The Burra Charter* and social inclusion', *International Journal of Heritage Studies* 12 (2006): 339–355.

West, Patricia. 'Uncovering and interpreting women's history at historic house Museums', in *Restoring Women's History through Historic Preservation*, edited Gail Dubrow and Jennifer Goodman. Baltimore: The John Hopkins University Press, 2003.

Worthing, Derick and Stephen Bond. *Managing Built Heritage: The Role of Cultural Significance*, Oxford: Blackwell, 2008.

Wright, Patrick. *On Living in an Old Country*, London: Verso, 1985.

Chapter 3

Historic Landscapes and the Recent Past: Whose History?

Peter Howard

On the wall of the secondary school close to my home someone has painted the words, 'In 1876, on this spot, nothing happened'. Such little conceits are now quite common, but they all share the same untruthfulness. It is certainly true that on that spot in 1876 there was no school or any other building, and the number of human feet to trample there would have been small indeed, although the ploughman would have passed that way. However, there would have been many more earthworms and cattle than at present. The length of history on that spot has been as great as any other spot on the Upper Permian formation – though we could go back before that also. All environments are equally historical. Historians may select some of them according to the very narrow limits of their discipline and call them 'historic', but the difference between 'historical' and 'historic' is an historian. The term 'historic environment' is very handy, immediately encapsulating the ownership of the concept, and the landscape itself, by a particular discipline. The phrase used by UNESCO to indicate a landscape where the works of human beings are of major significance is 'Cultural Landscape' and this seems to be widely accepted at international level, though not without some debate about its precise meaning. In Europe the word 'cultural' was omitted in the framing of the *European Landscape Convention* on the grounds that all European landscapes were cultural.[1] In England, however, such concepts become filtered through the Westminster system and English Heritage, and transformed into 'Historic Environment', as this would appear to be the only form of culture that can be recognized.[2]

Historical thinking invades at every opportunity. When writing my doctoral thesis in 1980, which examined the locations and subject matters of the landscape pictures exhibited at the Royal Academy from its inception to the present, it became obvious that the fashions there exhibited, at the home of what was once termed the 'negative of ambition' for *avant garde* artists,[3] were significantly out

---

1 Council of Europe, *European Landscape Convention*, 2000, http://conventions.coe.int/Treaty/en/Treaties/Html/176.htm.

2 See for example English Heritage, *Power of Place: The Future of the Historic Environment*, (London: English Heritage, 2000).

3 C. Harrison, *English Art and Modernism 1900–1939*, (Indianapolis: Indiana University Press, 1981), 17.

of alignment with those in the standard literature. P. Quennell dated the Romantic movement as essentially an eighteenth century one.[4] The Royal Academy figures demonstrated that the height of Romantic fashion was actually after 1850. The historian's understandable search for the source of ideas had wound the clock back almost a century. Although most of the disciplines concerned with heritage have broken out of their entrenched fields, there still remains, as Laurajane Smith has shown, an 'Authorized Heritage Discourse'.[5] She is an archaeologist who has broken out of the fascination for material objects, preferably old, but that still remains the focal point of most archaeological work. Geographers remain largely concerned with heritage as a place-centred phenomenon; many ecologists still have difficulty accepting the validity of human actions.

The concept of 'the heritage of the recent past' is one that from a landscape perspective is by no means clear. My own former property (see Figure 3.1) is a listed part of 'the heritage', built on a green-field site in 1843, but it still stands, so it is as much part of the landscape of 2007 as that of 1844. Heritage, virtually by definition, exists in the present and is 'present-centred'.[6]

Landscape is all of the same age, unless one is referring to geological time. If we recognize that Heritage is present-centred, and fulfils present needs, then surely the concept of 'the heritage of the recent past' means something quite different. The mediaeval heritage is that which was considered to be heritage in the Middle Ages – saints' relics for example.[7] So if the 'heritage of the recent past' has any meaning, it must refer to those items of nature/ monuments/sites/activities/ landscapes that were considered to be heritage in, for example, the 1930s or 1950s. Now that is a very interesting topic indeed. How has heritage changed? One clear referent for this is the change in the built heritage of England between Arthur Mee and Nikolaus Pevsner. Arthur Mee's *Devon*, for example was published in 1938, and the built heritage of each parish is represented largely by the parish church – about 79 per cent of the text concerns churches and cathedrals. Pevsner in the 1950s reduces this considerably, and increasing interest in vernacular buildings would reduce still further the dominance of the parish church in modern heritage writings.[8] In landscape terms, the thinking that led to the setting up of the English and Welsh National Parks was largely published in the 1930s and was itself based on the great shift in aesthetic taste of about 1870 which brought fells, fens, moors

---

4   P. Quennell, *Romantic England: Writing and Painting 1717–1851*, (New York: Macmillan, 1970).

5   L. Smith, *Uses of Heritage*, (London: Routledge, 2006).

6   B. Graham and P. Howard, 'Introduction: Heritage and Identity' in *The Ashgate Research Companion to Heritage and Identity*, edited by B. Graham and P. Howard, (Aldershot: Ashgate, 2008), 2.

7   See D.C. Harvey, 'The history of heritage', 19–36 in B. Graham and P. Howard, (see note 1).

8   A. Mee, *The King's England: Devon*, (London: Hodder & Stoughton, 1938); N. Pevsner, *The Buildings of England; South Devon*, (Harmondsworth: Penguin, 1952).

**Figure 3.1   Kerswell House, Devon. Built 1843. The former owner's view
includes the family's toys**

Source: © Peter Howard

and marshes onto the walls of our galleries; a taste that had trickled down to the
informed hiking public by the 1930s, and into political action a decade later.

The history of heritage is quite fascinating. Certainly by looking at landscape
one can see that the preferred landscape has changed dramatically over time, not
consistently but in major shifts, perhaps generational, and lasting about forty years.
However, only very few preferences actually disappear, as we find new interesting
landscapes, and dub them beautiful. The older tastes may be considered *passé* but
not actually ugly. Perhaps we need to pay very careful heed to those who are now
busy making the preferences of the future.[9]

Among the various trends that can apply to almost all the fields of heritage –
the move towards the vernacular for example, or the shift towards the conservation
of larger and larger areas, or habitats, is a trend towards democratic participation.
In the landscape field participation is firmly built into the European Landscape
Convention which the UK Government has now ratified. The position of the

---

9   P. Howard, *Landscapes: the Artists' Vision*, (London: Routledge, 1991).

academic or expert (and they may be different) is being questioned, even in some cases, such as the North Sea coast of Germany, actively disputed.[10]

So what role do we 'experts' play – or rather should we play? At present we play a very cunning game. We usually have much less money to expend to acquire the heritage we wish to control – much less than the owners of the resource, tourist companies, insiders, governments, (national and local) who can dispute control of heritage in the marketplace or the amphitheatre. The conspiracy theorist can imagine we experts sitting quietly behind Caesar (central Government), with our fancy academic journals, supported by the amenity societies that we dominate, and the 'quangoes'[11] which we staff, overriding all notions of democracy by quietly whispering what we want saved, and how. If Caesar does not listen, he may find himself denounced as culturally illiterate.

If we are to take democratic participation more seriously, there are four major roles for the experts which might appear valid, though the experts concerned may not always come from the same discipline. These are: invention, authentication, contextualization, and, education.

Invention

First, experts are partly responsible for inventing heritage, for deciding to which objects and activities that strange patina of the process of heritage is going to be applied. The history of heritage is full of individual sites and whole swathes of typologies of heritage that owe their origin to academic work. How many undergraduate dissertations and projects, let alone doctoral theses have 'discovered' new heritage sites – from henge monuments, to the gun emplacement on Exmouth beach that one of my students discovered. In the landscape field Judith Roberts described how we should be taking seriously the remains of 1930s suburban gardens.[12] Much earlier, John Swete, in Devon, applied the principles of the Picturesque, learned from Gilpin, to sites in his home county, thereby creating a map of visitable, and conservable, places that still remains partly intact.[13] The third-level Open University course on Modern Architecture where each student

---

10 W. Krauss, 'The natural and cultural landscape heritage of Northern Friesland', *International Journal of Heritage Studies*, 11/1, (2005): 39–52.

11 A quango is a 'Quasi autonomous non-governmental organization', such as Natural England, or English Heritage, which are largely Government-funded organizations, but may advise and even criticize Government.

12 J. Roberts, 'The gardens of Dunroamin; History and cultural values with specific reference to the gardens of the inter-War semi', *International Journal of Heritage Studies*, 1/4, (1996): 229–237.

13 J. Swete, 'Picturesque sketches of Devon', unpublished mss, 20 vols, (Exeter: Devon Record Office, ca 1810); W. Gilpin, *Observations on the River Wye and Several Parts of South Wales etc. Relative Chiefly to Picturesque Beauty*, (Richmond: Richmond Pub., 1782).

had to write a project on a significant twentieth-century building was a major source of knowledge, but also represented a proto-list for that period.

Shakespeare can help us with the study of the genesis of heritage; by misquoting Malvolio, we get 'Some things are born Heritage, some achieve Heritage, and some have Heritage thrust upon them'.[14] Things that are born heritage, that are intended as heritage from their origination, we usually call art, buildings for example, that were intended from their inception to be something other than merely functional. Other things reach this state by sheer survival, and it can be assumed that virtually any building from the seventeenth century or earlier will be carefully listed.[15] Most built heritage, though, has probably had heritage thrust upon it – it was just a cottage until Shakespeare was born in it. But actually it remained 'just a cottage' long after that, until someone recognized Shakespeare as an important writer. So Experts do a good deal of the thrusting and in so doing produce the accepted canon of what is considered heritage – the 'Authorized Heritage Discourse' that Smith has described.[16] Making such judgements is part of many professions; librarians for example have to consider the distinction between 'fiction' and 'literature', in the field of literary heritage. Of course, Experts are not the only ones who paint heritage onto things. As Raphael Samuel made clear, ordinary folk do it all the time.[17] They make collections of farm implement seats (see Figure 3.2), or perhaps Morris Minors. Of course, the heritage they invent soon appreciates in value and is sequestrated by the wealthy, at which time it may even attract the cultural capital that highly qualified experts can provide and become 'Authorized' by collecting into a museum.

In the landscape field different disciplines have invented heritage in different places, so we have a whole battery of designations which reflect the disciplinary authority – architectural historians play with listed buildings while archaeologists have scheduled monuments, then there are Regionally Important Geological Sites, archaeological parks, Sites of Special Scientific Interest for ecologists, or even National Parks – geographers always think big. So heritage values are culturally and historically constructed … of course they are. They are constructed by cultural historians, who are often anxious to let us believe that these values are immutable. They are not intrinsic values, except for the cost of the stone. The heritage value of a private house may even be a negative figure. Certainly there are some properties that cost more because they are designated heritage, but as Shipley makes clear in Canada, it cannot be taken for granted.[18]

---

14 Shakespeare, *Twelfth Night*, Act 2, Scene 5, Malvolio famously says 'some men are born great, some achieve greatness, and some have greatness thrust upon them'.

15 P. Howard, 'Asset formation and heritage policy' in *Cultural Tourism*, edited by J.M. Fladmark, 66–73, (London: Donhead, 1994).

16 L. Smith, *Uses of Heritage*, (London: Routledge, 2006).

17 R. Samuel, *Theatres of Memory: Volume 1: Past and Present in Contemporary Culture*, (London: Verso, 1994).

18 R. Shipley, 'Heritage designations and property values: Is there an effect?', *International Journal of Heritage Studies*, 6/1, (2000): 83–100.

**Figure 3.2   A collection of farm implement seats at the Devon County Show
                – the emphasis on the vernacular is a common trend**
Source: © Peter Howard

Authentication

Second, experts are required for purposes of authentication. This is a role with which
it is difficult to argue. Our educational and ethical background leads us to prefer the
genuine to the false, and in the heritage marketplace, the entire system of cultural
capital depends on authentication. Back in the 1960s the Hancock Natural History
Museum in Newcastle advertised the Mugwump, a bird that sat with its mug on one
side of the fence and its wump on the other. Somehow, even as an undergraduate,
though I smiled, yet I also felt that the museum was undercutting its own authority.
We must note that the expertise required for authentication is not that of the Heritage
Manager, but that of the ecologist, geologist, architectural historian, archaeologist,
whoever specializes in that field of heritage. Though that of course implies that we
know what field that is, and disciplines are very good at empire building. So we now
accept almost without question that churches belong to architects, not, for example,
to liturgists. Working in an art school as a geographer I was immensely impressed
by the connoisseur art historians' ability to deal with brushwork and dating, but
found their lack of interest in the landscape, or any other subject matter, to be quite
extraordinary. Similarly with photographic specialists, whose concerns used to focus
most precisely on discovering the name of the photographer, placing it in an artist's
oeuvre, and perhaps with the techniques used. Outside that narrow world, the interest
was almost exclusively on the subject matter of the photograph.

Landscape has had a strange journey in the last decades, and has moved from a largely scholarly concept, especially within the arts, to that proposed by the European Landscape Convention – as perceived by humans, and clearly perception is not merely visual. K. Olwig has made the distinction between the aristocratic landscape and the democratic *landschaft*.[19] Landscape has moved from the paintings of Friedrich, the expert and sensitized observer viewing places according to the guidebook, to that of Millet, where peasants stop a moment to cross themselves as the Angelus rings across the working fields, from observation to participation. In fact there seems to be scarcely any distinction between the new concept of landscape and the popular epistemology of place. If the academic concern with this new idea of landscape results in the elimination of the popular epistemology of place, or even worse its being subjected to academic analysis then the value of the amateur may be lost. Whimple, a village outside Exeter, recently was given a very substantial sum of money from the closure of a cider company, to found a heritage centre. The committee explained that the purpose of the centre was for the old villagers to explain to the new villagers what the heritage of Whimple was. The centre would be run by amateurs not because they could not afford professional help, but because it would be counter productive to that aim. Many ecomuseums are also founded on the advantage of the amateur.[20] But participation is not yet the norm, when selecting either landscapes or buildings for conservation status. In landscape surveys, now an essential part of the Environmental Impact Assessment required for major developments, the typical survey consists of one expert with a camera and a clipboard trying to convince us that his/her professional perspective carries more weight than that of a local resident.

All too often, the expert conducting a landscape survey for planning purposes will make no distinction between matters of fact and matters of opinion. The former might include the date of a building, an attribution to an architect, or the identification of a species, in all of which the local public are likely to accept expert advice. In matters of opinion, such as the significance of a building or the aesthetic appeal of a view, the public expect to have their say, especially when they are paying. The over-riding of public opinion in the latter field is not only undemocratic but calls into public question expert authority in matters of fact also.

The experts' strength is authority, so there is no surprise that their cry, when flushed from the cover of their 'quangoes' and Amenity Societies, is 'Authenticity'. This is

19   K. Olwig, *Landscape, Nature and the Body Politic*, (Madison: University of Wisconsin Press, 2002).

20   P. Davis, 'New museologies and the ecomuseum' in B. Graham B. and P. Howard, (see note 1).

rarely qualified. Which authenticity do you mean, for each discipline has its own
– the authenticity of the material? Of the context? Experience? Function? Site?[21]

Contextualization

The third task is to put the object (if indeed it is an object and not a part of the intangible
heritage) into context, although again one needs to specify which kind of context. The
geographical context might suggest this object was the only one in Europe, while the
context of scale suggested it was the third largest, and the historical context that it was
the second oldest. But it is within the contextual task that so many problems arise. The
contexts just mentioned still require the expertise of the parent discipline, but in the
landscape, the greater problem is to balance one discipline's statement of significance
against another. Of course, museums have to do this also. Some paintings will be hung
in the gallery devoted to local worthies rather than in the art gallery. But in the broader
landscape this decision is not the exception but the rule and will need the skills of
a broader heritage discipline to negotiate between several disciplinary agenda. The
decision whether to conserve the lichens or the stone on the cathedral is better not
left to an ecologist or an architect. Regrettably most curricula documents for Heritage
courses do not seem to focus on how one might balance such complicated conflicting
claims. Some, especially those with tourism in the course title, may be too ready to
take a decision based on forecasts of income generation. Perhaps it doesn't help that
so many heritage courses concern themselves almost exclusively with heritage sites
open to the public; it is so easy to be side-tracked by the tourist agenda, where the
sole function of the building or landscape is to be viewed as heritage. Students will all
learn about different stakeholders, but only rarely are experts themselves recognized
as stakeholders, or divided into their varied constituencies.[22] Should Glastonbury focus
on the abbey, the myth, the pop festival or the annual carnival? For most Glastonbury
people, it's obviously the latter. This issue of participation is exacerbated when dealing
with more recent buildings, or buildings with remembered uses.

Poltimore House is a ruined Grade 2* listed building (see Figure 3.3). Originally
built in the sixteenth century there were major changes and additions in every
century since. The Poltimore family (Bampfyldes) left in the 1920s and the place
became a girls' school, then a boys' school, then a private nursing home and finally
a hospital before abandonment in the 1980s. The Old Girls, the Old Boys, the
staff of the former hospital, those born in the nursing home, all have different
memories, and these memories result in quite different attitudes towards the
physical restoration. Former doctors suggest that the 1950's operating theatre (still
more or less intact) is of more historic value than the Tudor shell into which it was

---

21 G. Ashworth and P. Howard, *European Heritage Planning and Management*,
(Exeter: Intellect, 1999), 45.

22 W. Krauss, 'European landscapes: Heritage, participation and local communities'
in B. Graham and P. Howard, (see note 1), 425–438.

Figure 3.3   Poltimore House – here the media is a stakeholder, as the house portrays 'The Reichstag' in a World War II TV production
Source: © Peter Howard

inserted. Old Girls object most strongly to the proposed demolition of the grand staircase where they stood in lines waiting their punishment from an admittedly eccentric head teacher. Here neither the architectural historian, nor the family biographer can dictate the agenda.[23]

Education

The final task is education, both formal and, usually through interpretation, informal life-long learning. This is the area where academics and experts have the chance to ensure that a majority of the population will indeed trust their judgement, and will happily allow their hard-earned income to be diverted to the preservation of those things that the Experts nominate. It works very well, and local people indeed frequently do defer to the views of specialists. But how do we ensure that education does not spill over into indoctrination? Teachers during their training will learn the need for the ability of the taught to ask questions. Academics are taught

---

23   J. Hemming, *A Devon House: the Story of Poltimore*, (Plymouth: University of Plymouth Press, 2005).

the need for evidence. In interpretation, as in the pulpit, these conventions are dispensed with. The education strategy of the World Heritage Site of the Jurassic Coast of East Devon and Dorset for example tells us that 'Education is crucial for the successful implementation of all aspects of management', and Objective 6 is, 'disseminating a wide range of Jurassic Coast messages in a clear and effective way .... To ensure that the right messages are given'.[24] The right messages in this case are geological, as this site has clearly been intellectually colonized by that discipline. Would we accept such phrases if the sponsor were a political party?

Perhaps in the landscape field, if we have now reached the stage where our definition resonates with the popular epistemology, we may be able to produce an educational programme that is broader, and more holistic.[25] There can surely be no serious doubt of the need for education both in the significance of the objects classified as heritage and in the problems surrounding conservation and management. But, at least at the adult and interpretive end of the education spectrum, compulsory didacticism might be a questionable technique. A recent piece of research has demonstrated that ethnic minorities in an English cathedral city actually do not feel particularly excluded from the built heritage of England. Some prefer to opt in, and others to opt out, but they all object to recent attempts to be dragged within the embrace of national heritage.[26] They wish to retain choice, and few of our interpretation packages allow dissent, or allow a popular alternative. Having paid good money to enter the heritage site, we then find that the education package is compulsory. You are going to be given the lecture about these stones, even though your own interest is in the plants, or in just going for a swim. There are nearly always alternative histories. Griffiths showed in the Australian town of Beechworth that, to the locals, the most important building was the 1930s house once owned by the family that largely controlled the town, and failed to understand why the Experts were conserving some eighteenth century pigsties. The local agenda valued a building because of its inhabitants rather than its structure.[27]

Conclusion

Covert observation may be regarded as a research practice with some ethical problems, but a quiet hour or two spent in the organ loft of a typical parish church

---

24   Jurassic Coast, (undated), *The Strategy for Education*. Adopted consultancy report.

25   Dyer, A. (2007), 'Inspiration, enchantment and a sense of wonder ... Can a new paradigm in education bring nature and culture together again?' 86–97 in Howard, P. and Papayannis, T. (eds) *Natural Heritage: At the Interface of Nature and Culture*, (London: Routledge).

26   Shore, N., (2007), Whose heritage?: The construction of cultural built heritage in a pluralist, multicultural England, Unpublished Ph.D. Thesis, Newcastle-upon-Tyne: Newcastle University, 2007.

27   T. Griffiths, *Beechworth: An Australian Country Town and its Past*, (Melbourne: Greenhouse, 1987).

can be most instructive. There are indeed some visitors who have guidebook in hand and examine the architectural detail, but they are few. Another clear group is composed of those with family connections. But the large majority sit quietly in the nave, some praying but more simply quietly observing. They then perambulate the church and examine everything where people's names are mentioned – windows with donors' names, tombs, the list of incumbents, even the coffee rota. Frequently they will buy the history leaflet only on their way out, and perhaps return briefly to look at a particular item that is mentioned there. People connect with people. The Duke of Bedford found many years ago after opening Woburn Abbey that he and his wife were the most interesting exhibits, and people were more interested in their television set than the gold goblets.[28]

So we in the landscape field of heritage need to devise new methods to democratize the decisions as to which heritage is important. We shall need to understand that this will probably involve us with activities as much as objects, and with phenomena which date from a period within living memory. It does not necessarily mean more conservation, though certainly we need to expand our ideas about conservation systems, to include, for example, the county show as the means of conserving rural activities, but there are three other Cs as well. Some things can be collected, while for some sites, commemoration is the appropriate tool, although the American use of historic markers is not a major technique in UK. Finally copying may work. One of the most interesting undergraduate student dissertations I have seen was on the tribute band as a heritage phenomenon. This then leads us to consider the Renaissance as a heritage movement, aping the architecture of past times, the bricks-and-mortar equivalent of re-enactment.

We have to devise ways of introducing the other agenda – the local, the insider, the amateur, the private, and this will include encouraging insider groups to make decisions for themselves, and safeguard their own heritage. But should all local heritages be bruited from the roof-tops? Is it our job to strip out of every community the communal secrets that are such important glue? Smith recognized that the oral heritage of Australian Indigenous people should not be published in academic journals, and even that 'women's stories' should not be made available to men.[29] Do the local communities in this country have the same right? We need to get off our seat comfortably behind Caesar, and get down into the arena and dispute with others as to the future heritage. We experts have a valid view of what should be saved not least because of our real concern for future generations, even if by that we mean future cohorts of students in our own disciplines. We may be surprised both at how seriously we are taken, and by some of the landscapes that people want. The problems of population numbers and climate change also remind us that land is not a museum object, the primary function of which is self-

---

28 Duke of Bedford, *How to Run a Stately Home*, (London: André Deutsch, 1971).

29 L. Smith, A. Morgan, and A. van der Meer, 'Community-driven research in cultural heritage management: The Waanyi women's history project', *International Journal of Heritage Studies*, 9/1, (2003): 65–80.

preservation. The landscape, however historic, will have other demands made on it, some of them demands of life or death.

And, just as previous generations in the recent past had different ideas about what constituted their heritage, so will future generations. For example, regular visitors to exhibitions of landscape art will recognize the new preference for landscapes that are scruffy and unkempt, tumbledown sheds in allotment gardens, gnarled trees growing out of cracks in pavements, the mundane and often inhabited and used landscapes of everyday life. Pictures of Dartmoor now revel in abandoned tractors, overgrown hedges, corrugated iron sheds and bramble infested scrubland. As this new preference trickles down from the art schools to the population, how do we preserve untidiness?

Back in 1937 C.E.M. Joad wrote, 'The people's claim upon the English countryside is paramount ...; the people are not as yet ready to take up their claim without destroying that to which the claim is laid; .... the English countryside must be kept inviolate as a trust until such time as they are ready...'[30]

Are they ready yet?

Bibliography

Ashworth, G. and Howard, P. *European Heritage Planning and Management.* Exeter: Intellect, 1999.

Bedford, John, Duke of. *How to Run a Stately Home.* London: André Deutsch, 1971.

Council of Europe, *European Landscape Convention*, 2000, http://conventions. coe.int/Treaty/en/Treaties/Html/176.htm.

Davis, P. 'New museologies and the ecomuseum', in *The Ashgate Research Companion to Heritage and Identity* edited by B. Graham and P. Howard, 397– 414. Aldershot: Ashgate, 2008.

Dyer, A. 'Inspiration, enchantment and a sense of wonder: Can a new paradigm in education bring nature and culture together again' in *Natural Heritage: at the Interface of Nature and Culture* edited by P. Howard and T. Papayannis, 86–97. London: Routledge, 2007.

English Heritage. *Power of Place: The Future of the Historic Environment.* London: English Heritage, 2000.

Graham B. and P. Howard. 'Introduction: Heritage and Identity' in *The Ashgate Research Companion to Heritage and Identity* edited by B. Graham and P. Howard, 1–15. Aldershot: Ashgate, 2008.

Griffiths, T. *Beechworth: an Australian Country Town and its Past.* Melbourne: Greenhouse, 1987.

---

30   C.E.M. Joad, 'The People's Claim' in *Britain and the Beast,* edited by C. Williams-Ellis, C., (London: J.M. Dent and Sons, 1937), 64–85.

Harrison, C. *English Art and Modernism 1900–1939.* Indianapolis: Indiana University Press, 1981.

Harvey, D.C. 'The History of Heritage' in *The Ashgate Research Companion to Heritage and Identity* edited by B. Graham and P. Howard, 19–36. Aldershot: Ashgate, 2008.

Hemming. J. *A Devon House: the Story of Poltimore.* Plymouth: University of Plymouth Press, 2005.

Howard, P. *Landscapes: The Artists' Vision.* London: Routledge, 1991.

Howard, P. 'Asset formation and heritage policy' in *Cultural Tourism* edited by J.M. Fladmark, 66–73. London: Donhead, 1994.

Gilpin, W. *Observations on the River Wye and Several Parts of South Wales etc. relative Chiefly to Picturesque Beauty.* Richmond: Richmond Pub., 1782.

Joad, C.E.M. 'The People's Claim' in *Britain and the Beast* edited by C. Williams-Ellis, 64–85. London: J.M. Dent and Sons, 1937.

Jurassic Coast. *The Strategy for Education.* Adopted consultancy report, undated.

Krauss, W. 'The natural and cultural landscape heritage of Northern Friesland', *International Journal of Heritage Studies*, 11/1, (2005): 39–52.

Krauss, W. 'European landscapes: Heritage, participation and local communities' in *The Ashgate Research Companion to Heritage and Identity* edited by B. Graham and P. Howard, 425–438. Aldershot: Ashgate, 2008.

Mee, A. *The King's England: Devon.* London: Hodder & Stoughton, 1938.

Olwig, K. *Landscape, Nature and the Body Politic.* Madison: University of Wisconsin Press, 2002.

Pevsner, N. *The Buildings of England; South Devon.* Harmondsworth: Penguin, 1952.

Quennell, P. *Romantic England: Writing and Painting 1717–1851* New York: Macmillan, 1970.

Roberts, J. 'The gardens of Dunroamin; History and cultural values with specific reference to the gardens of the Inter-war semi', *International Journal of Heritage Studies*, 1/4 (1996): 229–237.

Samuel, R. *Theatres of Memory: Volume 1: Past and Present in Contemporary Culture.* London: Verso, 1994.

Shipley, R. 'Heritage designations and property values: Is there an effect?', *International Journal of Heritage Studies*, 6/1 (2000): 83–100.

Shore, N. *Whose heritage?: The Construction of Cultural Built Heritage in a Pluralist, Multicultural England*, Unpublished PhD Thesis, Newcastle-upon-Tyne: Newcastle University, 2007.

Smith, L. *Uses of Heritage.* London: Routledge, 2006.

Smith, L., Morgan, A. and van der Meer, A. 'Community-driven research in cultural heritage management: The Waanyi women's history project', *International Journal of Heritage Studies*, 9/1 (2003): 65–80.

Swete, J. 'Picturesque sketches of Devon'. unpublished mss, 20 vols. Exeter: Devon Record Office, ca 1810.

PART II
Cultural Landscapes

Chapter 4
Cultural Landscapes and Identity

Lisanne Gibson

Cultural landscapes are both artefacts within which we can trace past historical, social and cultural arrangements, and places which reflect fields of action confirming and negating contemporary arrangements of culture and society. No matter how picturesque they may be, cultural landscapes are spaces with real political, cultural and social effects on the present. Daily we wander past objects – memorials, monuments and so forth – often paying them no attention at all and yet these sometimes invisible 'signposts on cultural landscapes' can have significant impact both in terms of their presence and, as we shall see later, their absence.[1] The case of the memorial to the Kalkadoon and neighbouring Mitakoodi people in remote North West Queensland (QLD) in Australia is a sharp illustration of this point.

The Kalkadoon/Mitakoodi memorial, which features gentle reminders of Aboriginal land ownership, was erected as a bicentennial project in 1988 by long-time area resident and Cloncurry general practitioner Dr David Harvey Sutton. With the support of the Mitakoodi Aboriginal Corporation in Cloncurry, Sutton designed and financed a monument to draw public attention to Aboriginal history in the bicentenary year. The monument is located at Corella River on the Mount Isa to Cloncurry Highway at the purported tribal boundaries of the Kalkadoon and Mitakoodi peoples. It consists of a brick wall with images of the Aboriginal flag, Aboriginal warriors, spears, boomerang and plaques with poems on either side (see Figure 4.1). One of the central plaques reads:[2]

> You who pass by are now entering the ancient tribal lands of the Kalkadoon/Mitakoodi, dispossessed by the European; honour their name, be brother and sister to their descendants.

The overall effect of the arrangement is ambiguous. The monument's form, with its prominent flag, is something like a claim to territory, as is part of the inscription. Yet, as Chilla Bulbeck argues, other elements, the poems in particular, repeat the strategies that relegate Aboriginal culture to the past and gloss over both historical and contemporary struggles. The poems map out a version of Aboriginal 'progress', with descriptions of 'wild men' and 'brown Adam' who have adapted to the 'space

---

1   L. Gibson and J. Besley, *Monumental Queensland: Signposts on a Cultural Landscape*, (Brisbane: University of Queensland Press, 2004), 8–15.
2   L. Gibson and J. Besley, *Monumental Queensland*, 54.

**Figure 4.1   *Kalkadoon/Mitakoodi Memorial*, 1988. – the photo shows
                 Kalkadoon man John Brody**
Source: © Stephen Long

age' and whose 'children have been accepted as free, equal and valued members
of modern Australian society'.[3] There is no mention on the memorial of the 1884
battle between Europeans and Kalkadoon people at the nearby Battle Mountain
in which around 200 Aboriginal people died or indeed anything that points to the
distinctive experiences of Aboriginal people, a point that was further underscored
by the monument's unveiling by the Minister for Ethnic Affairs in the, at the time,
notoriously racist QLD State Government.[4]

---

3   C. Bulbeck, 'Aborigines, Memorials and the History of the Frontier' in *Packaging
the Past: Public Histories*, special issue of *Australian Historical Studies* edited by J. Rickard
and P. Spearritt, (1991): 169.

4   It may be that by 'unveiling' this monument the Government was attempting to
correct this general opinion as earlier in 1988 it had lost the watershed Mabo vs Queensland

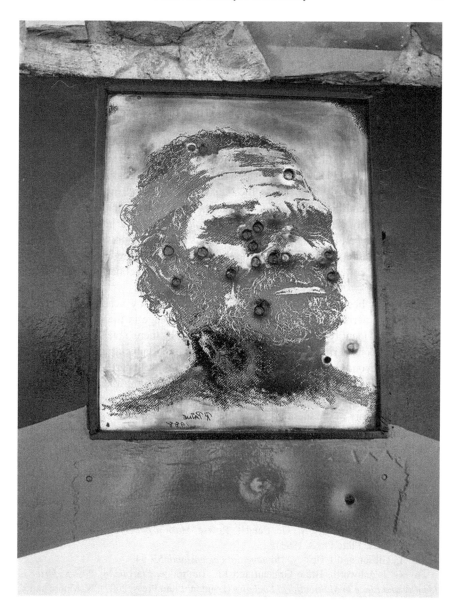

Figure 4.2 *Kalkadoon/Mitakoodi Memorial* – **detail showing bullet holes where the memorial has been shot, 1988**
Source: © Stephen Long

court case, in which the High Court of Australia decided that the *Queensland Coast Islands Declaratory Act*, which attempted to retrospectively abolish native title rights, was not valid

Despite the relatively conservative character of the monument it has been the subject of violent vandalism to the extent that explosives destroyed it in 1992. Although rebuilt, such attacks persist and demonstrate that positive representations of Aboriginality are not universally accepted in the context of contemporary Australian politics and society (see Figure 4.2).[5]

The Kalkadoon/Mitakoodi Memorial is not on either the State Heritage Register, neither is it protected under a local listing scheme. Would such protection make it less vulnerable to attack? The nearby Kalkadoon/Kalkatunga Memorial discussed later in this chapter, which specifically memorializes the conflict at Battle Mountain, is on the Queensland Heritage Register and has not been vandalized to this extent. But this could simply be because the Kalkadoon/Mitikoodi Memorial is laden with images including a large Aboriginal flag whereas the Kalkadoon/Kalkatunga Memorial simply carries a plaque. The key point here is that landscapes are not static in their meaning and are the site and focus of ongoing power struggles over what stories are represented. The representation of these stories in our landscapes has important cultural, political, social and perhaps even economic effects in the present.

There has been a great deal of consideration given by academics and, to a lesser extent, heritage practitioners to the role of heritage in the construction and articulation of identity. In the last five years in particular there have been a number of publications considering heritage as the locus of an important cultural politics which has effects beyond the cultural in the social and economic spheres.[6] This anchoring of considerations of heritage in relation to its concrete cultural, social and economic contexts and effects is not new,[7] but it is relatively recently that it has emerged as a significant feature of the literature. However, despite this, albeit recent, emphasis in the academic literature, the form of what counts as heritage (defined in terms of what gets *protected* as heritage) has changed little, as Barbara Bender comments:

---

according to the *Racial Discrimination Act 1975*. The subsequent Mabo vs Queensland case found, for the first time in Australian history, that native title existed in Australian common law, a finding that initiated the Commonwealth Government of Australia to develop the *Native Title Act*, 1993. See M. Goot and T. Rowse, *Make a Better Offer: The Politics of Mabo*, (Sydney: Pluto Press, 1994).

5   L. Gibson and J. Besley, *Monumental Queensland*, 53–54.

6   G. J. Ashworth, Brian Graham and J.E. Tunbridge, *Pluralizing Pasts: Heritage, Identity and Place in Multicultural Societies* (London: Pluto Press, 2007); N. Moore and Y. Whelan, *Heritage, Memory and the Politics of Identity: New Perspectives on the Cultural Landscape* (Aldershot: Ashgate, 2007); and, J. Littler and R. Naidoo, *The Politics of Heritage: The Legacies of Race* (Oxford: Routledge, 2005).

7   See for different instances: P. Wright, *On Living in an Old Country: The National Past in Contemporary Britain* (London: Verso, 1985); R. Hewison, *The Heritage Industry: Britain in a Climate of Change* (London: Methuen, 1987); R. Samuel, *Theatres of Memory. Volume 1: Past and Present in Contemporary Culture* (London: Verso, 1994); and, T. Bennett, *The Birth of the Museum: History, Theory, Politics* (Oxford: Routledge, 1995) especially chapters 9 and 10.

Apart from acting as the custodians and entrepreneurs of prehistoric monuments, English Heritage and the National Trust have focused on the landmarks of those with power and wealth, inscribed in an aesthetic that bypasses, as it has done for centuries, the labour that created the wealth. More recently the net has been cast wider, but, despite acquiring Victorian back-to-backs or derelict mills or mines, the presentation remains sanitized and romanticized, emphasizing local colour rather than the socio-economic conditions that generate both wealth and poverty, people's pain or their resistance.[8]

In the face of this critique, many writers have argued for a methodology based on consultation and consensus which will allow the management of heritage(s) which is plural and representative. But is 'heritage', at least in so far as it is articulated in contemporary policy and programmes, constitutively or even practically capable of being representative of a society in which deeply plural experiences and values are validated. Is consensus about what counts as heritage possible, or even desirable? In this chapter I want to tease out a variety of knotty issues around, in the first instance, the fabric focused logic underpinning heritage management and the implications of this for 'social value'. Second, I want to discuss the limits of heritage protection and the implications of this for the cultures and alternative histories of marginalized identities via a discussion of a detailed survey of 'outdoor cultural objects' undertaken on the Australian State of Queensland. In conclusion, I want to discuss critiques of heritage which propose consultation as a means of achieving processes of heritage management which are representative.

Valuing Cultural Landscapes

In this chapter I use the term cultural landscapes rather than 'historic environment' as I think that the former refers to places as not simply reflective of identities but as both reflective and *productive* of identities.[9] 'Historic environment' is a static understanding of landscape, as a canvas on which the evidences of history can be discerned. I prefer to think of landscape as a living environment in which signs and signifiers (whether historic or contemporary) and their meanings are constantly being remade. Unlike 'historic environment' as articulated in English Heritage documents as a canvas of the past, 'cultural landscapes' are a canvas of the present. Bender argues, 'Landscapes are thus polysemic, and not so much artefact as in

---

8   B. Bender, 'Contested landscapes: Medieval to present day', in *The Material Culture Reader*, ed. V. Buchli (Oxford: Berg, 2002), 141–174.

9   See the introduction to this book for a discussion of the recent policy history of this term, L. Gibson and J. Pendlebury, 'Introduction' in *Valuing Historic Environments*, (Farnham: Ashgate, 2009).

process of construction and reconstruction'.[10] Of course, this process is far from value neutral in the stories landscapes tell us about the past and the present. Denis Cosgrove, for instance, has shown how landscapes can be understood as part of a wider history of economy and society.[11] But landscapes most crucially tell us about the present; as Howard Morphy puts it 'the ancestral past is subject to the political map of the present'.[12] As discussed with the example of the Kalkadoon/Mitikoodi memorial, these signposts on landscapes are much more than the markers of stories without effect, rather, as Sharon Zukin argues, 'the symbolic economy of cultural meanings and representations implies real economic power'.[13] It is difficult to find any of this significance in relation to present day effects in the rather static terminology of 'historic environment'.

Another consequence of using the term 'cultural landscapes' rather than 'historic environment' is that it suggests an approach to heritage which understands it not as an object which is the static locus of some internal value, but as a process. Laurajane Smith has argued that 'there is no such thing as "heritage"'.[14] Rather, for Smith,

> heritage is heritage *because* it is subjected to the management and preservation/ conservation process, not because it simply *'is'*. This process does not just 'find' sites and places to manage and protect. It is itself a constitutive cultural process that identifies those things and places that can be given meaning and value as 'heritage', reflecting contemporary and cultural social values, debates and aspirations.[15]

There are a number of implications of Smith's statement which bear teasing out. In the first and most obvious sense, it follows from this position that there is nothing self-apparent or given about regimes of value and significance, rather these frameworks are specific to our particular social, cultural, economic and political contexts. Drawing on the anthropologist Marcel Mauss's famous proscription on the cultural and historical specificity of contemporary personhood, objects, buildings and places are 'formulated' as heritage 'only for us, amongst us'.[16] Perhaps even more significantly, Smith, drawing on the Foucaultian conception of

---

10   B. Bender, 'Introduction' in *Landscape: Politics and Perspectives*, ed. B. Bender, (Oxford: Berg, 1993), 2.

11   D. E. Cosgrove, *Social Formation and Symbolic Landscape* (Croom Helm, 1984), 1.

12   In Bender, 'Introduction' in *Landscape: Politics and Perspectives*, 2.

13   S. Zukin, 'Space and symbols in an age of decline', in *Re-presenting the City: Ethnicity, Capital and Cultural in the 21ˢᵗ Century Metropolis* ed. A.D. King, (Macmillan Press, 1996), 44.

14   L. Smith, *Uses of Heritage*, (Oxford: Routledge, 2006), 11.

15   Italics in original, ibid., 3.

16   M. Mauss, 'A category of the human mind: The notion of person; the notion of self' in *The Category of the Person: Anthropology, Philosophy, History*, eds. M. Carrithers, S. Collins and S. Lukes, (Cambridge: Cambridge University Press, 1985), 22.

'governmentality' as a productive process of power and knowledge, argues that it is the very 'governing' – the frameworks, instruments, policies and programmes – of heritage management which actually *produce* heritage.[17] Of course, as I have already begun to establish, this is far from a neutral process, in the case of heritage management, the 'for us and amongst us' has involved only limited sections of the population.

In the past 15 years the concepts of 'social value' and 'cultural significance' have gained ground as categories of value, enabling objects and places to be recognized when their significance is not primarily aesthetic or historical. The increasing emphasis on these categories is partly due to the criticism which many have identified that heritage management is too narrow, failing to reflect the breadth and depth of history, culture and society.[18] The concept of social value in heritage understands the locus of the value of objects and places as being within communities and people rather than in fabric whose meaning is identified by professionals. In a report on social value commissioned by the Australian Heritage Commission, Chris Johnston defines places with social value as those which can:

- provide a spiritual connection or traditional connection between past and present;
- tie the past and the present;
- help to give a disempowered group back its history;
- provide an essential reference point in a community's identity;
- loom large in the daily comings and goings of life;
- provide an essential community function that develops into an attachment; [and,]
- shape some aspect of community behaviour or attitudes.[19]

However, as we will see later in this chapter, social or cultural value is often difficult to establish for a variety of reasons. The primary reason is the fact that often the 'provenance' or the history of the social or cultural significance of an object has been lost or forgotten, a problem exacerbated when the object is perceived to be outside established heritage frameworks. For those arguing for social or cultural value it is the social or cultural significance attributed to the object or place

---

17 L. Smith, *Archaeological Theory and the Politics of Cultural Heritage*, (Oxford: Routledge, 2004).

18 C. Johnston, 'Social Values – Overview and Issues' in *People's Places: Identifying and Assessing Social Value for Communities*, (Canberra: Australian Heritage Commission, 1994a), 4.

19 C. Johnston, *What is Social Value?*, (Canberra Australian Government Printing Service, 1994b), 5. Johnston's definition was broadened by K. Winkworth, *Review of Existing Criteria for Assessing Significance Relevant to Moveable Heritage Collections and Objects*, (Canberra: Department of Communications and the Arts, 1998), to include collections and objects rather than only 'places'.

which is the source of value rather than the actual fabric of the object or place. The relative fragility of these intangible values poses significant challenges for heritage governance as we will see. Probably the most significant challenge, however, to arguments which attempt to give primacy to social and cultural value is the fact that, despite the rhetoric to the contrary, social and cultural value are still not widely accepted in actual practice as definitive categories of value in their own right.

Like other social and cultural policies and programmes contemporary heritage management must be articulated to values which are understood to define contemporary societies. In the case of both Australia and Britain these values are understood to be multi-ethnic and plural. From this it might seem that social and cultural value have become important in contemporary heritage policy. However, interrogation of key contemporary heritage management documents reveals that this importance is undermined and that in fact fabric focused, professionally defined (rather than community defined) value assessments remain the dominant paradigm of significance. We can demonstrate this by considering two key contemporary heritage policy documents.

Heritage policy and programmes in Australia are based on the Australia ICOMOS *Burra Charter*, 1999, which defines heritage significance in terms of 'aesthetic, historic, scientific or social value for past, present or future generations'.[20] *The Burra Charter* was first developed in 1979 in response to the *International Charter for the Conservation and Restoration of Monuments and Sites* (Venice 1964) and has been and still is very influential on heritage management instruments internationally. Its precepts and especially its definition of 'social value' and 'cultural significance' have been influential particularly in arguing for community inclusion in the processes of heritage management. The Charter's latest revision, in 1999, gave more emphasis to both the intangible aspects of heritage value and the importance of social and cultural value. It describes these as 'the recognition of less tangible aspects of cultural significance including those embodied in the use of heritage places, associations with a place and the meanings that places have for people'.[21] On this basis the Charter:

> recognizes the need to involve people in the decision-making process, particularly those that have strong associations with a place. These might be as patrons of

---

20 Australia ICOMOS, *The Burra Charter: The Australia ICOMOS Charter for Places of Cultural Significance*, (Victoria: Australia ICOMOS, 1999), Article 1.2, http://www.icomos.org/australia/burra.html (accessed 13/05/2008).

21 Ibid., Background.

the corner store, as workers in a factory or as community guardians of places of special value, whether of indigenous or European origin.[22]

Thus, the Charter appears to be assertively articulated to a community focused assessment of social value and cultural significance. This established understanding of the Charter has been challenged by Emma Waterton, Laurajane Smith and Gary Campbell who argue that 'the construction of terms such as fabric and cultural significance inherently contradicts attempts of social inclusion and community participation'.[23] Waterton et al. go on to show how, in contrast to its stated recognition of intangible meanings, the Charter's construction of the notion of fabric, 'assumes that cultural heritage is inherently fixed within, thus becoming physically manifested and subject to conservation, management and other technical practices'.[24] It follows from this that the locus of expertise implied by the Charter remains the heritage professional and not communities. As Waterton et al. argue '"Participants" are contrasted with "the experts", pushed into the role of beneficiaries, and thus made passive',[25] and therefore, 'the idea that the conservation values of experts might be just another set of cultural values is entirely absent in the discursive construction of the text, and for that matter all texts of this sort'.[26]

Even in the new English Heritage policies and guidance on the management of the historic environment *Conservation Principles*, the proposed conservation process places communication with people and communities as the second step after the evaluation of fabric (and the establishment of its significance) by experts.[27] This document sets out the framework on the basis of which the historic environment in England is managed by English Heritage.[28] However, analysis of the document reveals that, despite the community focus of English Heritage's rhetoric, the discourse which informs *Conservation Principles* places communication with people and communities as the second step after the evaluation of fabric (and the establishment of its significance) by experts. For instance, under the third proposed *Principle* on understanding the significance of places a process is described which establishes that 'it is necessary *first* to understand fabric'; *secondly*, the process will 'consider

---

22 Ibid.

23 E. Waterton, L. Smith and G. Campbell, 'The Utility of Discourse Analysis to Heritage Studies: *The Burra Charter* and Social Inclusion', *International Journal of Heritage Studies*, 12, 4 (2006): 347.

24 Ibid., 348.

25 Ibid., 350.

26 Ibid., 349.

27 English Heritage, *Conservation Principles, Policies and Guidance for the Sustainable Management of the Historic Environment* (London: English Heritage, 2008).

28 Ibid.

who values the place and why they do so'.[29] Thus fabric and expertise always come before value as established by non-expert communities. This is despite the fact that the second *Principle* proposes that 'Everyone should have the opportunity to contribute to understanding and sustaining the historic environment. Judgements about the values of places and decisions about their future should be made in ways that are accessible, inclusive and informed'.[30] Yet *Principle* 2.3 states that 'Experts would use their knowledge and skills to help and encourage others to learn about, value and care for the historic environment'.[31] As Waterton et al. identified in *The Burra Charter*, so to, the English Heritage *Conservation Principles* are defined by a fabric focus which looks to expertise as the locus for the establishment of meaning and value. What are the implications of this fabric focus for an actual cultural landscape?

Cultural Heritage Mapping

I want to explore some of these issues by discussing research I undertook with Joanna Besley from 2000 to 2003 in Queensland (QLD). The 'Public Art and Heritage' project was funded by the Australian Research Council and also received some funding from the Public Art Agency and Cultural Heritage Branch of the QLD State Government. Between 1999 and 2007 the Public Art Agency was responsible for QLD's 'per cent for art' scheme which funded public art, defined broadly to include built art forms such as statuary, installations, building design elements, and so forth, it could also include ephemeral cultural forms such as festivals, dance, and so forth (although in practice it rarely did). The Cultural Heritage Branch is responsible for managing the QLD Heritage Register and the State's heritage programmes. The 'Public Art and Heritage' project aimed to investigate outdoor cultural heritage and public art in the State, its management, and the extent to which it was possible to establish strategic policy linkages between the two departments and the policies and programmes they managed.

The research needs to be viewed in the context of the State's introduction of an *Integrated Planning Act* in 1997 which required QLD local governments to undertake both cultural heritage mapping and cultural mapping in order to inform the development of cultural plans for the management of all local cultural resources (the receipt of state Government arts funding was dependant on the development of these plans).[32] On the face of it, this impetus was based on the widely accepted best practice cultural planning aims of recognizing the multiple agencies across Government responsible, in various ways, for cultural programmes

---

29   Ibid., 21, italics added.

30   Ibid., 20.

31   Ibid.

32   Queensland Government, *Integrated Planning Act, 1997*, (Brisbane: Queensland Government, 1997).

and resources.[33] It was believed that identification of all cultural resources which could then be managed using a 'whole of Government' approach both at State and local government levels would result in a more pragmatic management of the State's cultural assets and would also result in more inclusive policies and programmes due to a more integrated system for managing the diversity of assets in the State.[34] As it turned out by 2002, five years after the introduction of the *Act*, only 45 per cent of the 128 local councils and 20 per cent of the 32 indigenous councils had undertaken the mapping.[35]

The cultural heritage mapping process was based on what were considered to be best practice methods of public consultation directed at achieving a plural identification of cultural heritage. However, the process could only enable a limited pluralization of response. This was so for three primary reasons:

1. Local governments engaged in the cultural heritage mapping process by following the Queensland Government's *Guidelines for Cultural Heritage Management*.[36] The consultation procedure defined in this document provided a framework of nine historical themes which were significant to the States' history against which a particular localities heritage was to be mapped. The guidelines suggested that councils could introduce additional sub-themes to deal with the specific local history of an area; however in practice this was rarely done.

2. Public consultation was a required part of the cultural heritage mapping process; however, the requirement for consultation was interpreted very differently by the councils. Our survey of the organizational bodies responsible for the management of heritage – 128 local councils and 28 indigenous councils – found that of the small number of councils who had conducted cultural heritage studies or surveys, only 50 per cent included public consultation as a component.[37] In addition, on further analysis we found that activities defined as 'public consultation' included appointing a committee of heritage experts and appointing a member of the public to sit on it.[38] Other activities defined as 'public consultation' most commonly took the form of public meetings, meetings with local specialist groups such as

---

33   See for instance Australian Local Government Association, *Making the Connections: Towards Integrated Local Area Planning*, (Canberra: Australian Local Government Association, 1992), and D. Grogan, C. Mercer and D. Engwicht, *The Cultural Planning Handbook: An Essential Australian Guide*, (Sydney: Allen and Unwin, 1995).

34   L. Gibson, *The Governance of Heritage Significance and the Protection of Public Art*, (Brisbane: Queensland Government, 2004).

35   Ibid., 25.

36   Cultural Heritage Branch, *Guidelines for Cultural Heritage Management*, (Brisbane: Environmental Protection Agency, Queensland Government, 2001).

37   L. Gibson, *The Governance of Heritage Significance*, 22.

38   Ibid.

historical societies and meetings with individual members of the community, known to be experts.[39] Such activities are unlikely to encourage participation by the general public or a representative understanding of cultural significance. Problems with public consultation are well documented;[40] not least the significant resources and time required to undertake consultation which aims for more than token community engagement. While the State Government required public consultation to take place in the process of cultural mapping, they provided no funds to enable this. Little wonder then, that such limited consultation actually occurred.

3. The third reason why this seemingly best practice model for the establishment of a pluralized definition of cultural heritage resources could not result in a genuinely representative understanding was the very constitution of the process which lay at the heart of *Guidelines for Cultural Heritage Management*. The heritage process as defined by this document provided a limited framework which was biased to events understood to be part of an already established State history and which was limited to tangible objects and places. Having established this limited lens through which heritage could be defined, the document also established the role of the expert as the transmitter of this framework and thus ultimate arbiter of what came to be understood as culturally significant.

Signposts on a Cultural Landscape

While the difficulties with achieving a representative understanding of heritage value are significant, the designation and protection of some objects, buildings and places and not others has considerable discursive and practical consequences. I have discussed some of the theoretical coordinates of this acknowledgement of the cultural, economic, political and social consequences of 'heritage' in the first section of this chapter and in the introduction to this book.[41] Here I want to discuss through a range of examples the importance of objects in local cultural landscapes and in relation to the cultural landscapes of states and nations.

As part of the 'Public Art and Heritage' project we surveyed the existence, provenance and heritage management of a category of objects we called 'outdoor cultural objects' throughout the State of QLD. We defined outdoor cultural objects

---

39   Cultural Heritage Branch, *Guidelines for Cultural Heritage Management*, (Brisbane: Environmental Protection Agency, Queensland Government, 2001).

40   See for instance P. Bourdieu, 'Public opinion does not exist' in *Communication and Class Struggle*, eds. A. Mattelart and S. Siegelaub, (International General, 1979) and T. Bennett, 'Making culture, changing society: The perspective of "culture studies"', *Cultural Studies* 21, 4–5, (2007): 626.

41   L. Gibson and J. Pendlebury, 'Introduction' in this volume.

as objects intended to have a representative, memorial or symbolic function and which were located in places easily accessible to the public. This was taken to include sculptural art, monuments, memorials, mosaics, murals, tympaneaum, building badges (with a sculptural element), outdoor advertising objects (such as 'Big Things'), sundials, and fountains. We documented over 200 objects, the vast majority of which were not on either the State Heritage Register or local cultural heritage maps. Many of the objects had common characteristics which were in contrast to our expectations. We assumed that most of the objects would be installed by Government programmes to provide reminders of Empire, or models of worthy citizens to be emulated. In contrast the majority of the objects we discovered were installed by community subscription or community groups. So while they were not listed in the Heritage Register or on local cultural heritage maps, the fact that many of these objects were there as a result of community efforts demonstrated that the local stories that these objects represented were significant to that community. The specificity of these objects was particularly poignant when we looked at the overall survey of the cultural landscape of QLD and the story communicated through the outdoor cultural objects on it. The story told by the 'official' cultural heritage landscape of QLD (those things protected through the State Heritage Register or through local listings or mapping) is a story in which women, Indigenous Australians and the majority of QLD's ethnic communities, including communities, such as the Chinese, who have lived there for more than 200 years, are close to absent. In contrast, the stories told by the objects we surveyed in the 'unofficial' cultural landscape of QLD were comparatively diverse, although as we will see the absences are still significant.

Monuments and memorials which tell of the contribution of women, and working women in particular, to the story of QLD are almost absent from QLD's cultural landscape. Where women are represented in the landscape, they are represented as symbolic of particular roles or ideals, such as Queen Victoria, or the *Weeping Mother War Memorial* (see Figure 4.3).

We found only two objects in the entire State which represented women as workers, both of which were memorials to Sister Elisabeth Kenny.[42] The monuments in Toowoomba (see Figure 4.4) and Nobby were installed by community subscription but are on neither the State Heritage Register nor the local cultural heritage maps. It was a women's organization (in this case the Country Women's Association, responsible for the installation of many of the memorials in QLD) that was the instigator of the Sister Kenny Memorial Garden in Nobby. Sister Kenny's methods of medical treatment were recognized overseas before they were accepted in Australia, particularly in America, where she was depicted as a heroine in a Hollywood movie, *Sister Kenny* (1946), about her life. Kenny invented a treatment for polio based on the application of hot compresses and massage. This form of treatment was in direct contradiction to the established medical procedures of the time, which involved the application of splints and immobilization of the limbs.

---

42 L. Gibson and J. Besley, *Monumental Queensland*, 142–43.

Figure 4.3   *The Weeping Mother Memorial* **by Frank Williams and Co., 1922**
Source: © Richard Stringer

Brisbane doctors ridiculed her methods at a Government-sponsored demonstration and the report of a Royal Commission conducted by leading QLD doctors in

Figure 4.4   The Sister Kenny memorial sundial in Toowoomba, QLD
Source: © Toowoomba Historical Society

1937 damned her methods.[43] Over time, her technique of gradual and gentle rehabilitation based on mobilization of the limbs became recognized by the medical establishment as the best method for the treatment of polio.

The absence of women in the landscape and moreover the lack of 'official' recognition of those markers which do exist is not exclusive to QLD. 'The Unusual Monuments Project', an Australia-wide survey that sought to catalogue all the monuments that depicted labour history, women's history, and the history of Indigenous Australians, found that there were very few monuments to the work of women.[44] This survey found that despite the fact that 'the female form abounds in many monuments... it is not women who are represented but rather abstract ideals... Less common are the monuments which celebrate dead or living females'.[45] As we found through our survey of the State of QLD, Bulbeck found that 'There are more monuments to women in Australia than a reading of the various registers of Australian monuments... would suggest'.[46] There are a number

---

43   Ibid.
44   C. Bulbeck, 'Women of Substance: The Depiction of Women in Australian Monuments', *Hecate* XVII, II, (1992): 8- 29.
45   Ibid., 8.
46   Ibid.

of reasons Bulbeck identifies to account for this: the lack of signposts – tourist brochures, official and local histories in which women 'do not find a place';[47] and, as we found also, the relative material insignificance of memorial's to women which tend to be plaques, small cairns and such like, rather than imposing statues located in busy places.[48]

The opening of the 'Women in Australia's Working History Project' in 2002 at the Australian Workers Heritage Centre provided an opportunity, not yet realized, to address the lack of commemorative objects to women workers in Queensland's cultural and urban landscapes. The project seeks to correct the imbalance and to celebrate the lives, work and contribution of Australian women past and present.[49] The project is the first occasion on which women's contribution as workers in the home, in the community and in the paid workforce is collectively recognized in a permanent tribute on a national scale. However, as an illustration of Bulbeck's point about the relative material invisibility of women's history compare this small static exhibition with *The Ringer* a monumental sculpture of a stockman installed just down the road at the Australian Outback Heritage Centre (see Figure 4.5). In this very material comparison we find reasserted the hegemonic view that the story of QLD (and Australia) is a story almost exclusively of heroic white working men.

Perhaps unsurprising given the well-known biases of Australian history is the relative absence of outdoor cultural objects marking the histories of Indigenous Australians. Prior to 1970, outdoor cultural objects mostly reflected orthodox versions of Australian history.[50] These objects utilized European ways of identifying achievements and marking the landscape, such as the erection of plaques, statues, cairns and other commemorative strategies. The material form of these objects and the attitudes reflected in their inscriptions are, therefore, an inevitably selective and Eurocentric account. Thus, in many of these objects Indigenous Australians are depicted as helpmeets to a larger European story.

With the emergence of other perspectives on Australian history and increasing public and, to a limited extent, legal recognition of the demands of Aboriginal people, discursive spaces are gradually being made for alternative versions of Australian stories, although conflicts about dissonant national, regional and local stories are an object of vigorous and ongoing public debate.[51] There are not as yet many outdoor

---

47   Ibid.

48   Ibid.

49   Australian Workers Heritage Centre, *Opening Celebration of Stage One of the Women in Australia's Working History Project*, (Barcaldine: Australian Workers Heritage Centre, 2002).

50   C. Bulbeck, 'Aborigines, memorials and the history of the frontier' in *Packaging the Past. Public Histories*, special issue of *Australian Historical Studies* edited by J. Rickard and P. Spearritt, (1991), 169.

51   See in relation to the Australian Museum, for instance, D. Casey, 'Museums as agents for social and political change' in *Museums and their Communities*, ed. S. Watson, (London: Routledge, 2007).

Figure 4.5   *The Ringer* by Eddie Hackman, 1988
Source: © Stockman's Hall of Fame

cultural objects that directly represent or embody these challenges to traditional historical interpretations. However, there are a number that assert Indigenous identity and culture or utilize Indigenous forms of aesthetic expression, for instance, the artworks by Indigenous artists Virginia Jones and Ron Hurley in memory of Oodgeroo

Figure 4.6	*Oodgeroo, the Woman of the Paperbark*, **by Virginia Jones, 1996**
Source: © Courtney Pedersen

Noonuccal, the Brisbane based poet and activist known as Kath Walker[52] (see Figure 4.6) and 'Reconciliation Path' by Indigenous artists Bianca Beetson and Paula Payne in the Boondall Wetlands.[53] All of these works were commissioned by public art programmes run by a University in the former and a local council in the latter case.

The importance of cultural programmes, including public art and heritage programmes which enable alternative stories to be signposted in the landscape is clear when we have some idea of the absence of Indigenous Australian history from the QLD landscape. For instance, despite the prevalence of war memorials in the State, Indigenous Australians are significant by their absence despite their active participation in the wars in which Australia has fought. Anzac Square in Brisbane is the location for the eternal memorial flame and as such is the physical and discursive centre for all the State's war memorials. This Square has memorials representing all participants in the wars including recently installed memorials to women and war animals; however, as yet Indigenous Australians are absent. In fact, we found only one memorial which mentions Indigenous Australian participation in the world wars and this was installed in 1992 by the Yugambah people of the Gold Coast region to commemorate their people who fought in the two world wars, Malaya, Borneo and Vietnam.[54]

Only one memorial in Queensland directly acknowledges the history of conflict related to the invasion of Australia by Europeans. Like many other plaques mounted on stone cairns, this one commemorates a centenary – one hundred years since the slaughter of Kalkatungu people at Battle Mountain and 'one hundred years of

---

52	L. Gibson and J. Besley, *Monumental Queensland*, 210.
53	Ibid., 227.
54	Ibid., 115.

survival' of their descendants. Erected by the Kalkatungu/ Kalkadoon people in 1984 at the Kajabbi bush pub north-west of Cloncurry and twenty kilometres from Battle Mountain itself, the memorial explicitly links the events of the past with the political issues of the present. Unveiled by Charles Perkins, an Aboriginal activist and senior public servant, and George Thorpe, a Kalkadoon elder, the plaque reads in part:[55]

> This obelisk is in memorial to the Kalkatungu tribe, who during September 1884 fought one of Australia's historical battles of resistance against a para-military force of European settlers and the Queensland Native Mounted Police at a place known today as Battle Mountain – 20 kms south west of Kajabbi.

> The spirit of the Kalkatungu tribe never died at battle, but remains intact and alive today within the Kalkadoon Tribal Council. 'Kalkatungu heritage is not the name behind the person, but the person behind the name'.

The message is a direct affirmation of contemporary Aboriginal social and political structures as well as a statement of pride in the past and present collective actions of Aboriginal people. The cairn not only commemorates the battle but was also timed to celebrate the founding of the newly formed Kalkadoon Tribal Council. The cultural significance of Battle Mountain as the site of an event that symbolizes Aboriginal resistance to colonial invasion is acknowledged by the inclusion of this place on the *Register of the National Estate*.[56]

As I argued in the introduction to this chapter the representation and official recognition (through heritage protection) of some stories and not others has concrete and real effects on the everyday. Some of QLD's other significant communities are totally absent from local heritage maps, cultural registers and heritage programmes. For instance, we could find no representations of the very significant Vietnamese community, which in Brisbane we found particularly surprising. Perhaps even more surprising was the absence of representations of stories of the various Chinese communities from around the State, particularly given the relatively long history of some of these communities.[57]

Despite the capacity to protect heritage on the basis of social value which is enshrined in QLD's heritage legislation and part of the framework for the local cultural heritage mapping, a glance through the State Register and the local

---

55   Ibid., 51–53.

56   Ibid.

57   As judged by language spoken in the home the Chinese community is the largest and the Vietnamese community the third largest foreign language speaking community in the State. The top five languages spoken at home in Queensland were Chinese (1.0%), Italian (0.7%), Vietnamese (0.4%), German (0.4%) and Greek (0.3%) according to Queensland Government, Office of Economic and Statistical Research, *Queensland Characteristics: A State Comparison (Census 2001 Bulletin no. 15)*, http://www.oesr.qld.gov.au/queensland-by-theme/demography/population-characteristics/bulletins/qld-characteristics-state-comp-c01/qld-characteristics-state-comp-c01.shtml, (accessed 17/07/08).

heritage maps shows that the objects and places that are managed as heritage tend to have particular physical characteristics, which allow them to be understood as architecturally or aesthetically significant. This presents a particular problem for the State's objects which are symbolic of diverse stories for two reasons. Firstly, many of the unlisted objects we found had been installed or developed by community fund raising, as many of the communities in QLD are small and are not wealthy, communities are only able to draw on very limited resources. The lack of funds available means that objects commissioned through community subscription are often not materially significant, at least in accepted aesthetic or architectural terms. In the second place, in many cases due to lack of management, the provenance of these objects has been lost. This loss of provenance is a significant factor which militates against such objects being listed or at least locally managed.

As the example of the Kalkadoon/ Mitakoodi Memorial discussed in the introduction to this chapter makes clear, despite their lack of formal recognition as 'heritage', objects such as these are far from insignificant. A recent survey of Australian attitudes to the past *'Australians and the Past'* challenges the 'assumption that Australian's knowledge of history arises from formal teaching and officially endorsed accounts of the past'.[58] Rather, in a survey involving 350 telephone and 150 face-to-face interviews Paul Ashton and Paula Hamilton found that the most important medium for historical narrative was objects and places.[59] Commemoration and memorialization were found to 'have close associations with historical objects and places'.[60] However, objects and places were important not for a value internal to the object or place, such as an aesthetic value, but as vehicles for remembering or as 'a driving or anchoring force in narratives'.[61] Above all though Ashton and Hamilton found that history and what was valued as history was a highly personal and individual construct. Given the articulation of 'heritage' to unifying identity constructs such as 'national' how can we develop a process for valuing 'heritage' which is attendant to the radically individual nature of people's historical stories, which would manage dissonant versions of heritage, but also be practical in its applications. Or is such a thing impossible?

Heritage Representation: Is 'Community' the Answer?

There are volumes of literature dissecting specific heritage management or interpretation programmes and critiquing them for not being representative.[62]

58　P. Ashton and P. Hamilton, 'At home with the past: Background and initial findings from the National Survey', *Australian Cultural History* 22, (2003), 10.

59　Ibid., 13.

60　Ibid., 23.

61　Ibid., 13.

62　See for instance of an Australian case study K. Markwell, D. Stevenson and D. Rowe, 'Footsteps and memories: Interpreting an Australian urban landscape through

Such critiques usually propose two related requirements to enable a 'heritage' which is plural and representative. First, a high level of self awareness on the part of the heritage process, programme, and practitioner is proposed such that the process/ programme/ practitioner is not influenced by the hegemonic power/ knowledge relations which inform and limit 'heritage'. As Emma Waterton puts it in relation to archaeologists we must be aware 'of those knowledge systems that have allowed archaeologists to distance themselves from the political and social acts of archaeology'.[63] Markwell, Stevenson and Rowe argue that 'in spite of its aims and legitimating rhetoric, much cultural planning [including heritage management] is occurring without the active involvement of local communities, and with little regard for the specificities and complex histories of place'.[64] In Markwell, Stevenson and Rowe's critique we find the second element purportedly required for the achievement of a 'representative heritage' namely community consultation. In the first part of this chapter I identified in brief some of the *practical* challenges with community consultation. Here I want to explore some of the theoretical problems with the notion of community consultation. This is not with the aim of questioning whether the participation of peoples and communities should be a key part of any cultural programme or policy which involves public resources. On the contrary, it is precisely because I believe public participation should be a much more central feature of the heritage process that I want here in conclusion to engage briefly with some of theoretical and methodological issues involved.

According to the French sociologist Pierre Bourdieu 'public opinion does not exist'.[65] Bourdieu argues that public opinion as evidenced by surveys can never be a representation of unfettered public opinion as it is always framed and moulded by the framework of the survey, and the capacity that publics have to participate.[66] As Tony Bennett argues social science research methods designed to identify community opinion actually produce 'public opinion' 'as a new agent and mode of action on the social'.[67] Andrea Witcomb has identified a similar effect in relation to heritage and museums. Witcomb concurs with Bennett's analysis that 'community' is a construct as much of social science surveys, as of museum and heritage 'community' programmes.[68] Nevertheless, she argues that Bennett goes too far in

---

thematic walking tours', *International Journal of Heritage Studies* 10, 5, (2004): 457–473 and for a UK case study see E. Waterton, 2005, 'Whose sense of place? Reconciling archaeological perspectives with community values: Cultural landscapes in England', *International Journal of Heritage Studies* 11, 4, (2005): 309–325.

63   E. Waterton, 'Whose sense of place?', 319.
64   K. Markwell, D. Stevenson and D. Rowe, 'Footsteps and memories', 458.
65   P. Bourdieu, 'Public opinion does not exist', 1979.
66   Ibid.
67   T. Bennett, 'Making culture, changing society', 626.
68   A. Witcomb, *Re-Imagining the Museum: Beyond the Mausoleum*, (London: Routledge, 2003), 81.

his analysis and thus he cannot account for the *agency* of communities. Witcomb proposes instead to understand museums, heritage organisations and practitioners as themselves communities of practice. It then becomes possible to think of working in a dialogical communication model between multiple communities both practice and non-practice. This is in contrast to transmission based models of communication where consultation follows expertise, as illustrated by heritage documents such as English Heritage's *Conservation Principles* discussed earlier. However, processes of 'dialogue' are far from free of power/ knowledge relations. As Jeffrey Minson describes it,

> talk about the ethical-political value of participatory democracy almost invariably makes reference to its *ideal* forms of expression (principles, ends, values, prefigurative models). The result: endless stand-offs and unstable compromises between champions of the participatory ideal and "realists"'.[69]

Minson proposes avoiding these 'intellectual and sentimental cul-de-sacs' by attending to the practical and pragmatic *procedures* for participation. For Minson a democratic heritage process would involve a system which equipped citizens with the capacities necessary for participation. This is not to discount Witcomb's analysis of communities of practice involved in dialogue with 'non-practice' communities but rather to add another dimension to this through the recognition that enabling the agency of peoples and 'communities', is only possible within frameworks of action and by utilizing particular techniques and capacities *which are not innate.* It is crucial that we are clear eyed in our understanding of what Bennett terms the 'ontological politics' at play,[70] which in the interaction of cultural knowledge's and social sciences establishes an ontological triangulation of heritage, identity and representation. It is in its very essentialism that the power and the danger of this equivalence lies. If we want to aid the formulation of landscapes which speak more broadly than is currently the case, and given the affects of these on the contemporary everyday, it is surely imperative that we do so. What this might involve is attendance to the micro, mundane and practical programmes which aim to equip peoples with the capacities necessary for various kinds of participation.

In the case of the 'outdoor cultural objects' we discovered in QLD this would perhaps involve introducing a flexibility which is impossible to conceive within the current expert led 'protection of fabric' based model of heritage management. Rather as architecture and heritage studies academic Kim Dovey argues:

> Educating communities to research and defend their places of value is easier to justify than the determination that such places are to be protected by law against development. Such an approach avoids some of the dilemmas ... in that

---

    69   J. Minson, 'The participatory imperative' in *Questions of Conduct: Sexual Harassment, Citizenship, Government*, (London: Macmillan, 1993), 190.

    70   T. Bennett, 'Making culture, changing society', 626.

it does not measure, define, judge or paralyse places of social value. Rather it empowers and enables people to define themselves and places as part of the general development of democratic social life.[71]

The implications of this for the heritage sector might include programmes which equip peoples with the capacities to articulate the things and places they value as heritage or even indeed to create new markers of past or present heritage. Establishing ways to empower communities and peoples in this way is the challenge not only for the heritage sector but for all cultural policy.

Acknowledgements

Thanks to Richard Sandell for very useful and detailed comments on an earlier version of this paper. Thanks also to John Pendlebury and the participants in three colloquia on Value and the Historic Environment held at the University's of Leicester and Newcastle during 2007 for the informative and thought provoking discussions which have informed my thinking here.

Bibliography

Ashton, P., and Hamilton, P. 'At home with the past: Background and initial findings from the National Survey'. *Australian Cultural History* 22, (2003): 5–30.

Ashworth, G. J., Graham, Brian and Tunbridge, J. E. *Pluralising Pasts: Heritage, Identity and Place in Multicultural Societies.* London: Pluto Press, 2007.

Australia ICOMOS, *The Burra Charter: The Australia ICOMOS Charter for Places of Cultural Significance.* Victoria: Australia ICOMOS, 1999. http://www.icomos.org/australia/burra.html.

Australian Local Government Association, *Making the Connections: Towards Integrated Local Area Planning.* Canberra: Australian Local Government Association, 1992.

Australian Workers Heritage Centre. *Opening Celebration of Stage One of the Women in Australia's Working History Project.* Barcaldine: Australian Workers Heritage Centre, 2002.

Bender, B. 'Contested landscapes: Medieval to present day' in *The Material Culture Reader*, edited by V. Buchli, 141–174. Oxford: Berg, 2002.

Bender, B. 'Introduction' in *Landscape: Politics and Perspectives*, edited by B. Bender, 1–17. Oxford: Berg, 1993.

71   K. Dovey, 'On registering place experience' in *People's Places: Identifying and Assessing Social Value for Communities*, (Canberra: Australian Heritage Commission, 1994), 33.

Bennett, T. 'Making culture, changing society: The perspective of "culture studies"'. *Cultural Studies* 21, 4–5 (2007): 610- 629.

Bennett, T. *The Birth of the Museum: History, Theory, Politics*. Oxford: Routledge, 1995.

Bourdieu, P. 'Public opinion does not exist' in *Communication and Class Struggle*, edited by A. Mattelart and S. Siegelaub, New York: International General, 1979.

Bulbeck, C. 'Women of substance: The depiction of women in Australian monuments'. *Hecate* XVII, II, (1992): 8–29.

Bulbeck, C. 'Aborigines, memorials and the history of the frontier' in *Packaging the Past. Public Histories*, special issue of *Australian Historical Studies* edited by J. Rickard and P. Spearritt, (1991): 168–178.

Casey, D. 'Museums as agents for social and political change' in *Museums and their Communities* edited by S. Watson, 292–299. London: Routledge, 2007.

Cosgrove, Denis E. *Social Formation and Symbolic Landscape*. Croom Helm, 1984.

Cultural Heritage Branch. *Guidelines for Cultural Heritage Management*. Brisbane: Environmental Protection Agency, Queensland Government, 2001.

Davison, G., 'The meanings of "heritage"' in *A Heritage Handbook*, edited by G. Davison and C. McConville, 1–13. Sydney: Allen & Unwin, 1991.

Dovey, K. 'On registering place experience' in *People's Places: Identifying and Assessing Social Value for Communities*. Canberra: Australian Heritage Commission, 1994.

English Heritage, *Conservation Principles, Policies and Guidance for the Sustainable Management of the Historic Environment*. London: English Heritage, 2008.

Gibson, L. *The Governance of Heritage Significance and the Protection of Public Art*. Brisbane: Queensland Government, 2004. Available at http://hdl.handle.net/2381/130.

Gibson, L. and Besley, J. *Monumental Queensland: Signposts on a Cultural Landscape*. Brisbane: University of Queensland Press, 2004.

Gibson, L. and Pendlebury, J. 'Introduction' in *Valuing Historic Environments* edited by L. Gibson and J. Pendlebury. Farnham: Ashgate, 2009.

Goot, M. and Rowse, T. *Make a Better Offer: The Politics of Mabo*. Sydney: Pluto Press, 1994.

Grogan, D., C. Mercer and D. Engwicht, *The Cultural Planning Handbook: An Essential Australian Guide*. Sydney: Allen and Unwin, 1995.

Hewison, R. *The Heritage Industry: Britain in a Climate of Change*. London: Methuen, 1987.

Johnston, C. 'Social values – overview and issues', in *People's Places: Identifying and Assessing Social Value for Communities*. Canberra: Australian Heritage Commission, 1994a.

Johnston, C. *What is Social Value?*. Canberra: Australian Government Printing Service, 1994b.

Laidley Shire Council, 2000, *Draft Heritage Study Brief*, 10 July, Laidley Shire Council.

Littler, J. and Naidoo, R. *The Politics of Heritage: The Legacies of Race*. Oxford: Routledge, 2005.

Lowenthal, D. *The Heritage Crusade and the Spoils of History*. Cambridge: Cambridge University Press, 1998.

Lowenthal, D. *The Past is a Foreign Country*. Cambridge: Cambridge University Press, 1985.

Markwell, K., Stevenson, D., and Rowe, D. 'Footsteps and memories: Interpreting an Australian urban landscape through thematic walking tours'. *International Journal of Heritage Studies* 10, 5 (2004): 457–473.

Mauss, M. 'A Category of the human mind: The notion of person; the notion of self' in *The Category of the Person: Anthropology, Philosophy, History*, edited by M. Carrithers, S. Collins and S. Lukes, 1–25. Cambridge: Cambridge University Press, 1985.

Minson, J. 'The Participatory Imperative'. *Questions of Conduct: Sexual Harassment, Citizenship, Government*, 190–253. London: Macmillan, 1993.

Moore, N. and Whelan, Y. *Heritage, Memory and the Politics of Identity: New Perspectives on the Cultural Landscape*. Aldershot: Ashgate, 2007

Queensland Government, Office of Economic and Statistical Research. *Queensland Characteristics: A State Comparison (Census 2001 Bulletin no. 15)*. http://www. oesr.qld.gov.au/queensland-by-theme/demography/population-characteristics/ bulletins/qld-characteristics-state-comp-c01/qld-characteristics-state-comp-c01.shtml (accessed 04/06/08).

Queensland Government. *Integrated Planning Act, 1997*. Brisbane: Queensland Government, 1997.

Samuel, R. *Theatres of Memory. Volume 1: Past and Present in Contemporary Culture*. London: Verso, 1994.

Smith, L. *Archaeological Theory and the Politics of Cultural Heritage*. Oxford: Routledge, 2004.

Smith, L. *Uses of Heritage*. Oxford: Routledge, 2006.

Waterton, E. 'Whose sense of place? Reconciling archaeological perspectives with community values: Cultural landscapes in England'. *International Journal of Heritage Studies* 11, 4 (2005): 309–325.

Waterton, E., Smith, L., and Campbell, G. 'The utility of discourse analysis to heritage studies: *The Burra Charter* and social inclusion'. *International Journal of Heritage Studies* 12, 4 (2006): 339–355.

Winkworth, K. *Review of Existing Criteria for Assessing Significance Relevant to Moveable Heritage Collections and Objects*. Canberra: Department of Communications and the Arts, 1998.

Witcomb, A. *Re-Imagining the Museum: Beyond the Mausoleum*. London: Routledge, 2003.

Wright, P. *On Living in an Old Country: The National Past in Contemporary Britain*. London: Verso, 1985.

Zukin, S. 'Space and symbols in an age of decline', in *Re-presenting the City: Ethnicity, Capital and Cultural in the 21ˢᵗ Century Metropolis* edited by A.D. King, 43–59, London: Macmillan Press, 1996.

Chapter 5

Being Autocentric: Towards Symmetry in Heritage Management Practices

John Schofield

This chapter will consider cultural heritage and landscape, and how people engage with them. A particular focus will be those places people feel attached to, often evoking more personal meanings and values, amounting to a more 'intimate engagement'[1] than state-led mechanisms for heritage management currently allow. I will argue that these intimate engagements are often with mundane or 'everyday' places, as opposed to those that are deemed to be special, and those typically afforded statutory protection. One might imagine for example: a gateway through which one last saw a close friend or relative; a burial place; a house where we grew up; or simply somewhere closely familiar whose loss or removal would be at best disorientating and at worst traumatic. This loss might involve a building, or an element of landscape character – a particular field boundary or streetscape; even a particular type of field boundary or streetscape which evokes a sense of place, of local-ness, of comfort and familiarity. Occasionally these particular intimate engagements converge, often where memorable events occurred that many people experienced together: anti-nuclear protests at Greenham Common; or environmental protests on Twyford Down; or amongst Aboriginal people, places where wild food resources were gathered, or places of ritual; or perhaps a settlement that was destroyed, either in conflict or development, uniting its former residents in grief.

Whether converged or isolated, it is these places where the strongest sense of attachment tends often to be felt. It is specifically places like these that I am thinking of when I use the term 'everyday archaeology': it is the archaeology of us, and of our everyday lives; where memory and some physical manifestation of place come together.[2]

This chapter will take the form of an exploration of some everyday aspects of the historic environment (here given its broad, holistic, inclusive definition, implying continuity in space, not a discontinuous landscape populated by particular and 'special' things with constraint lines drawn around them), aspects

---

1   John Schofield, 'Intimate engagements: Art, heritage and experience at the "place ballet"', *International Journal of the Arts in Society* (2007): 105–114.

2   One might also coin the phrase 'everyday' or 'popular heritage', as being something broader and all-embracing, the latter mirroring the distinction between 'high' and 'popular culture'.

which are valued by the people that inhabit them, or have done so in the past. It will set this exploration within the wider context of potentially millions of other similar, more personal, explorations, and will begin to assess why these intimate engagements matter and how they can be made more widely available to those who want to share in them, through new and increasingly mobile technologies for example. The essay constitutes an alternative view, set alongside and sometimes in direct contradiction to the official view of how we value, present, and manage the historic environment.

Symmetrical Practices

> Symmetry isn't just for snowflakes and mirrors. Look deeper and you find it rules the universe.[3]

This contribution takes as its starting point heritage management practices as they exist today amongst western societies, as well as the profound changes in social and political context that are causing us to rethink and reassess these practices; to question their currency, suitability and relevance.[4] Yet, notwithstanding current changes such as those to planning and heritage protection in England and Wales, many of the laws that offer protection to cultural heritage sites, it could be argued, remain grounded in late nineteenth-century legislation and ideals. Are these laws still fit-for purpose, some will ask? Can this and other aspects of heritage management practice be modernized in a way that reflects recent changes to society and governance? My contention here is that it can, and that a symmetrical approach provides the framework for doing so. That said, heritage protection reform in England and Wales could go some way to achieving the symmetry that I refer to.

But what is a symmetrical approach, precisely, and how might this manifest itself in a realignment of socially relevant heritage management practices? Numerous definitions have been offered recently for symmetrical archaeology, such as a 'new ecology packed with things, mixed with humans and companion species and which prioritizes the multi-temporal and multi-sensorial qualities, the multiplicity, of the material world'.[5] Of its slogans and key concepts, Witmore goes on to list the following:

- Archaeology begins with mixtures, not bifurcations.
- There is always a variety of agencies whether human or otherwise.

---

3   M. du Sautoy, 'Grand designs', *New Scientist* 198 (2008): 38.

4   For example in England and Wales where heritage protection reform is, at the time of writing, creating a new system that offers greater clarification, partnership and participation than existed previously.

5   Chris Witmore, 'Symmetrical archaeology: excerpts of a manifesto', *World Archaeology* 39, 4 (2007): 547.

- There is more to understanding than meaning.
- Change is spawned out of fluctuating relations between entities, not of event revolutions in linear temporality.
- The past is not exclusively past.
- Humanity begins with things.

Symmetry implies multiple views and perspectives in other words. Take Witmore's point about time. He explains how we know landscape to be a 'complex aggregate mixture of disparate eras, events, achievements which have a durable trace', and that 'in articulating this ensemble archaeologists can treat time as the sorter and situate each component in relation to another'.[6] He emphasizes the point also about the continuity of the archaeological record, that pasts are not wholly past, thus opening up a whole new world of *ta archaia*, recognizing how modern (like ancient) things are equally gatherings of achievements from various times and places. The landscape, and the material culture it contains, is layered as in a palimpsest – layers are written, erased and rewritten, while the processes of the palimpsest lend themselves to entropy, decay and stratification. The landscape has changed and will continue to change as we and our successors mould and shape it with our actions. Change is not something uniquely past as some heritage management practices have perhaps tried to ensure. But as Witmore says, this only partially takes us down the very long road towards understanding the nature of time. What about points of connection, proximity and action between various pasts? What of the subtle pleats and folds in the fabric of time?[7]

Simply by taking these comparatively straightforward and largely uncontested components of archaeological practice, the passage of time and the palimpsest, archaeology becomes by definition more 'everyday', more relevant and less 'exceptional'; and heritage can embrace the material trace of these everyday activities, activities that on one level are mundane and ordinary, yet can become special in their ordinariness. We do not need the State for this to happen (though it can helpfully perform an enabling role). Communities can do it for themselves, and they will do so if any sense of attachment and ownership exists at all. Tuan describes 'the cult of the past', noting how organized historic preservation has little relevance to the connection of people to place:

> The state of rootedness is essentially subconscious; it means that a people have come to identify themselves with a particular locality, to feel that it is their home and the home of their ancestors. [And] a truly rooted community may have shrines and monuments, but it is unlikely to have museums and societies for the preservation of the past.[8]

---

6   Ibid.

7   Ibid: 556.

8   Yi-Fu Tuan, *Space and Place: The Perspective of Experience* (Minneapolis: University of Minnesota Press, 1977), 194, 198.

Such a symmetrical approach to heritage management practice forms the basis of this chapter, recognizing that: a) change happens; and that b) a multiplicity of views will exist about the past, the present and about future change. But let us begin, simply, by taking one example from one of the proponents of symmetrical archaeology to clearly illustrate its application in a heritage context: 'the ethical imperative to incorporate other voices in archaeological practice at sites deemed germane to many other interests and activities that may not subscribe to an archaeological canon of knowledge and manifestation'.[9] Using the example of local and Mexican visitors to Teotihuacan (Mexico), a symmetrical approach demonstrates how 'the past is with us every day, acting upon us and questioning our claimed humanist and modernist 'liberation' from the uncivilized worlds of things and pasts.[10] Webmoor goes on to note that:

> In 'taking things seriously', an examination of the 'heritage' of Teotihuacan, or public and stakeholder involvement with these sites of the archaeological imagination, reveals a set of four primary associations inter-linked and stabilized thanks to the work of things themselves: spiritual or new age practices, diversion, economics, archaeology. The core of these associations forms what is normally taken for heritage, but a heritage extracted from the circulation of humans and things through other primary associations. Tracing these associations out from a single site draws awareness to local and national identity politics, indigenous and stakeholder beliefs and rights, and the overall archaeological 'footprint' upon local and regional communities. But it also grants a 'voice' to a far greater assembly: to both humans *and* things.[11]

A symmetrical approach therefore provides a window to the everyday and the commonplace, events and occurrences which shape our lives and social practices and give our lives pattern and meaning. As we will see, cultural heritage is no longer confined to protected and demarcated sites and places that have special status and value – places over which local communities have little influence, and which are the sole concern of official heritage agencies and Government bodies. Heritage also concerns the everyday, and a symmetrical approach can ensure this is accommodated in our working practices. Further, through a symmetrical approach we can celebrate the everyday, recognizing how we shape it and how it shapes us, thus introducing an element of performativity into our heritage procedures and practices.

To further investigate these everyday places I will begin by contextualizing the heritage practices that exist today, and briefly consider how cultural heritage is becoming increasingly multi-vocal, continuous, holistic and relevant. I will then

---

9   Tim Webmoor, 'What about 'one more turn after the social' in archaeological reasoning?', *World Archaeology* 39, 4 (2007): 573.

10   Ibid., 575.

11   Ibid., 574–5.

consider how these principles can influence heritage management practices and society's access to and place in the historic environment.

Heritage Now

Since heritage legislation was introduced to the UK in 1882 and even since the first buildings' protection in 1932, society has changed irrevocably.[12] Within our communities we have become socially and culturally more diverse, with politics and politicians gradually attuning to that diversity, in aspects of education and social policy for example. Further, people are increasingly aware (and made-aware) of their heritage, and encouraged to participate in it; to research it, understand it, and be supportive of a desire to preserve it. Increasingly people do support the heritage, as a MORI opinion poll that accompanied English Heritage's (1999) *Power of Place* document revealed (this is discussed further below). Yet still archaeological sites are protected according to many of the same criteria as were used over a century ago, and buildings by the same measure as in the 1930s, criteria that reflect a particular view and very specific perceptions of what constitutes value. Broadly speaking these criteria are universal, as are their shortcomings.

Cultural diversity is key here, as is the realization that it is increasingly unacceptable to simply support a predominant view to which all members of society are expected to conform. In another *Power of Place*,[13] Dolores Hayden noted that of Los Angeles' 3.5 million people in 1990, just under 40 per cent were Hispanic, 37 per cent white, 13 per cent black, just under 10 per cent Asian American or Pacific Islander and 0.5 per cent Native American. Further, she noted how Los Angeles was the largest Mexican, Armenian, Filipino, Salvadoran and Guatamalan city in the world, the third largest Canadian city, and had the largest Japanese, Iranian, Cambodian and Gypsy communities in the United States. Hollywood High School housed students who spoke 35 native languages. Los Angeles' population has always been diverse. However city biographies and landmarks fail to reflect that diversity. They each favour a small minority of white, male landowners, bankers, business and political leaders. In fact 98 per cent of Los Angeles's designated cultural-historic landmarks are Anglo-American, and only 4 per cent were associated with any aspect of women's history. So, as Dolores Hayden said, three-quarters of the current population must find its public, collective past in a small fraction of the city's monuments, or live with someone else's choices about the city's history. The major ethnic groups that have always

---

12   Roger M. Thomas, 'Archaeology and authority in the twenty-first century', in *The Heritage Reader*, edited by Graham Fairclough, Rodney Harrison, John H. Jameson Jnr. and John Schofield (London: Routledge, 2008), 139–148.

13   Dolores Hayden, *The Power of Place: Urban Landscapes as Public History* (Cambridge [Mass.]: The MIT Press, 1995).

been part of the city have been dispossessed. And the new immigrants have every reason to be confused, she said.

We can see the same in London. In the area of Tower Hamlets over 90 per cent of the resident population is Bengalee. There are many listed buildings in this area – buildings that are protected by statute on the advice of English Heritage and its predecessors. Separately, the Bengalee residents have clear and strongly-held views on the built environment and what they value about it.[14] Yet the official 'expert' view of what matters bears no relation at all to the locally-held opinions of residents. They value their religious houses, community centres, markets and meeting areas; yet officially it is the historic churches and works by renowned architects that matter more. Gard'ner's analysis helpfully concludes with recommendations, suggesting ways in which locally held views can be given weight and status. He describes the benefit of 'local lists' for example, based on a Geographical Information System (GIS) and periodically reassessed and reviewed by the community through meaningful consultation. Local lists offer a form of designation suitable for recognizing more recent buildings and those of local rather than national interest. Another option is the revision to the boundaries of conservation areas, ensuring the inclusion of places and buildings valued by the local community. As Gard'ner points out, the politically motivated nature of the Bengalee community has enabled it to lobby for the extension of existing conservation areas and initiate the creation of new ones, some with a distinctly British Bengalee identity. This has also encouraged the retention and imaginative reuse of historic urban fabric, and has enhanced the amenity and streetscape through partnership funding. There is the erection of plaques and other historical markers, as recommended in paragraph 6.15 of PPG15 (*Planning and the Historic Environment*). By using these in conjunction with self-initiated community history projects, historical markers can be used to identify sites and events of importance to groups like the Japanese–American community in Los Angeles.[15] But perhaps of greatest relevance here is the degree to which character assessment can be used as a way of measuring heritage value in a non-hierarchical way. As Gard'ner puts it:

> The value of recognizing and protecting the character of an area is echoed in recommendation 13 of [English Heritage's] *Power of Place*, which proposes the strengthening of the conservation area designation through reducing permitted development rights. Characterization studies of the type currently being promoted by English Heritage offer a mechanism for including the more intangible or ephemeral values associated with an area and facilitate a more

---

14   James M. Gard'ner, 'Heritage protection and social inclusion: a case study from the Bangladeshi community on East London', *International Journal of Heritage Studies* 10, 1 (2004): 75–92.

15   Gail Dubrow, 'Feminism and multicultural perspectives on preservation planning', in *Making the Invisible Visible: A Multicultural Planning History*, edited by L. Sandercock (Berkeley: University of California Press, 1998), 57–77.

socially inclusive understanding of the range of values that minority groups may place on their heritage.[16]

A symmetrical approach would involve the dispossessed and the confused in deciding what matters most, and what to do with it. And this can work – as has been demonstrated in Australia where migrant communities are being encouraged to understand and give value to 'their' heritage places. I'll return to this example later. It worked also at Winston Street (Washington DC), where Ulf Hannerz's early work *Soulside* explored the significance of a narrow, one-way ghetto street whose resident community he divided into two contrasting but overlapping and interacting groups: the 'mainstreamers' and the 'swingers'. Hannerz showed how the neighbourhood and its institutions – bootlegging, numbers gaming, storefront churches, 'running mouths' on front steps and street corners, music, and the pervasive nature of 'soul' – formed a culturally cohesive whole. The neighbourhood is important to its residents, he said, because of 'the significance of a ghetto-specific culture to community integration'.[17]

Let us now address briefly the concept of time. Traditionally there has been a chronological separation between past and present, with heritage experts only taking a view on the significance of those artefacts, ancient monuments and historic buildings 'from antiquity' – *ta archaia*. When the first *Schedule of Ancient Monuments* was drawn up in 1882, all twenty-four sites were prehistoric and monumental. With time the Schedule expanded to include Roman sites, then medieval, post-medieval and more recently industrial and modern military places such as Greenham Common.

Gradually, the separation between past and present has been eroded therefore, and the present has become past (Figure 5.1). As archaeology has become more concerned with recent material culture, so the heritage sector has responded by commissioning research that can inform conservation decisions about it. Which monuments of the Cold War era do we most wish to keep for example? How should we balance protecting historic landscape character with the inevitable processes of change and creation that form part of that character? Here, specifically, significant progress has been made in recent years and symmetry is clearly in evidence. Still there is bias towards earlier remains. Still the older things are the more valuable they often appear to be. Still change is viewed by some with distaste and suspicion. Yet for archaeologists change is the thing we understand best of all: things change – we can't prevent that. What we can do as archaeologists and heritage practitioners is to recognize the process and the impacts of change, and manage that process accordingly, working with it rather than against it. The past in the present, yes, but recognizing also that how we manage change now affects the past that communities will encounter in the future.

---

16   Gard'ner, (see note 14).

17   Ulf Hannerz, *Soulside: Inquiries into Ghetto Culture and Community* (Stockholm: Almqvist and Wiksell, 1969), 158.

Figure 5.1   Symmetry (1). The past is not always past – we constantly change, re-use, alter and adjust our surroundings in a way that allows us to realize the relevance of continuity of change, and to therefore juxtapose old and new; here the extremes of a Lower Palaeolithic landscape at Boxgrove (West Sussex, UK) and a Cold War landscape at Greenham Common (West Berkshire, UK)

Source: © Dave Hooley (Boxgrove); John Schofield (Greenham Common)

Once, planners would consider the impact of major developments on the historic environment by viewing options – for a road scheme, a major housing development – in relation to a map of recorded archaeological sites and historic buildings; the archaeological *constraints* in other words. Put simply, planners and developers would take the view that by simply avoiding mapped constraints, all would be well. But as we all know those dots are fairly meaningless: they record interventions and events, and thus bear only the vaguest of relations to past activity and thus to measuring true environmental impact. The technology now exists to take a more holistic view. In the UK and Europe, historic landscape characterization offers a broader view of landscape character, being the historic processes that have given particular areas of landscape their distinctive characteristics.[18] These can be large upland areas, fields with their associated settlements, or they can be areas of market towns for example, or an area of terraced housing in a former industrial centre that has fallen on hard times. Current proposals exist in England to build thousands of new homes in the south-east of the country in the next two to three decades. Where should these houses go? By simply plotting out all records of known archaeological sites (notably scheduled monuments and sites on local Historic Environment Records), listed buildings and other constraints, developers and planners could simply produce a scheme that avoids these known and 'special' places as far as possible, and which takes other factors such as access, waste management and so on into account. But characterization can enable a much more sophisticated and helpful dialogue between developers, planners, local communities and the heritage sector. Based on a GIS, characterization documents the historic fabric of the countryside *as a whole*, allowing us all to understand at the touch of a mouse how widespread well-preserved medieval field patterns are for example, how much boundary loss has occurred already, how sensitive to change the various character areas are, and what form and density of settlement is characteristic of the areas concerned. Some areas may be characterized by small dispersed hamlets and farmsteads, with little change over the past 300 years. Other areas may have been populated in nucleated settlements, and seen much more expansion and change. These latter areas may be better able to accommodate further changes than the areas of dispersed settlement, for instance.

This approach has an obvious application, and provides the basis for dialogue, not just involving planners, developers and heritage practitioners, but politicians and local communities as well. And it promotes a clear and important message: that the historic environment is not just about point data. It exists everywhere, all around us (Figure 5.2).

And the 'everywhere' does truly matter. People value their historic environment, especially when their own special places, their landscapes, are given recognition, even if only by being 'polygonized' on a map. There is a danger that heritage – as officially sanctioned and undertaken – will commodify and package things – places, objects, reports on interventions – in a way that separates them off from society.

18   See www.english-heritage.org.uk/characterisation for examples and applications.

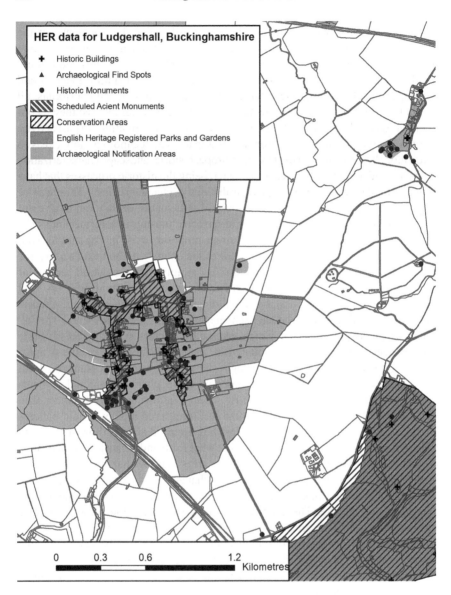

Figure 5.2   Symmetry (2). What sort of landscape exists only in the special places that someone has identified, mapped or designated? The landscape is a continuous, holistic thing, constantly on the move, perceived by (any) people, and the result of interaction between natural and/or human factors. Here the discontinuity of dots-on-maps is contrasted with the continuity of historic landscape characterization (HLC). Only one of these can claim symmetry of approach and integrity

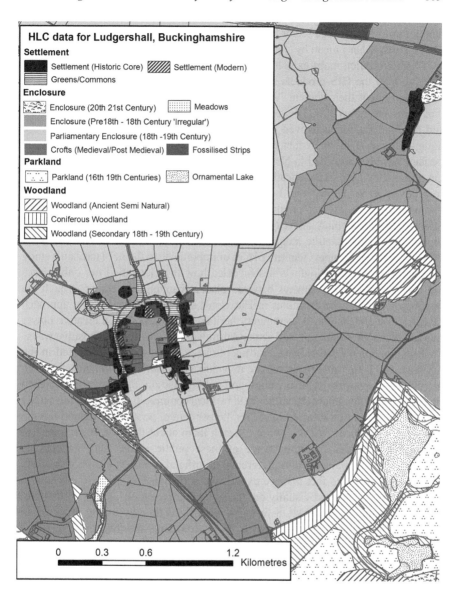

Figure 5.2   Symmetry (2) continued
Source: Both images © Crown Copyright. All rights reserved. Buckinghamshire County Council, 100021529 2008

Material things in other words become reports, photographs, management plans. We need to ensure that these things are relevant to people in the form of everyday places that remain or become accessible, for example through GIS and the Internet.

And for us in the heritage sector to be seen to be taking an interest is important too. This is partly what I mean by 'being autocentric', to which I shall return.

Characterization maps exist now covering most of England and the technology exists to take the GIS products around local communities, and engage people in discussion about their content. We can encourage local communities to generate their own layers on the GIS, cognitive maps of their own values, perceptions and personal experiences, even (perhaps especially) where those experiences are traumatic ones. Similar in fact to work in Australia where Aboriginal Women's heritage has been recorded in New South Wales. Women tell their stories based on the places they have experienced, and events that happened there. As one woman said, 'everywhere I look here holds a memory. Even out at the Headlands and Swimming Creek. It doesn't matter where you go, there are memories.'[19] Harrison refers to this as mapping landscape biographies or 'mapping attachment', in which topographic maps, satellite images and aerial photographs are used to create base maps on which informants identify places of importance to them.[20] Harrison's study of pastoralists in New South Wales revealed how they identified not only places but also the trails and pathways between places, mapping movement through the landscape. In all of the examples in Harrison's study, an archaeological approach reveals close attachment to community heritage, and introduces a creative power that can reconcile Aboriginal and settler Australians in powerful and positive ways.

It is perhaps no coincidence that people have become more interested in the heritage, and better able to appreciate its value and meaning, commensurate with a growing recognition of local values, and willingness amongst heritage bodies to understand and document the values held by local community groups. In the MORI poll of public opinion referred to earlier, 96 per cent of people in England felt the historic environment was important to teach them about the past. A symmetrical approach recognizes the importance of these public values and perceptions, alongside (and perhaps even in preference to) official or expert views of the past. It also recognizes the fact that we experience and engage with these places in a number of ways, not just visually but through other sensations, and with thought, feeling and intuition. ICOMOS recently recognized this, by taking more of an interest in intangible heritage – cuisine, music, traditions and folklore.[21]

In his *Environmental Aesthetics* (1996)[22] Porteous makes the useful distinction between an autocentric (or subject-centred) mode of perception which combines sensory quality and pleasure, concerning how people feel; and an allocentric mode

19   Kath Schilling, *Aboriginal Women's Heritage: Nambucca* (NSW National Parks and Wildlife Service, 2003).

20   Rodney Harrison, *Shared Landscapes: Archaeologies of Attachment and the Pastoral Industry in New South Wales* (UNSW Press: Department of Environment and Conservation [NSW], 2004).

21   http://www.unesco.org/culture/ich/index.php.

22   J. Douglas Porteous, *Environmental Aesthetics: Ideas, Politics And Planning* (London and New York: Routledge, 1996).

Table 5.1   Autocentric and allocentric senses (after Porteous 1996: 31)

Autocentric	Allocentric
'hot'	'cool'
physical	intellectual
primitive	sophisticated
expressed de novo on each occasion	easier to recall and to communicate
more 'actual'	more detached
close range	distanced
basic in children	develop with maturation

which is concerned more with objectification and knowledge (see Table 5.1). We can read this distinction in two ways. First, that, as Porteous says, Western people tend to prize the cool, detached and intellectual sense which effectively distances them from their environment, and that with such distance it becomes rather too easy to regard nature and environment as a series of objects worthy only of disregard and exploitation. Second, we can see in this the distinction between the attitudes of those closely familiar with their local area (or a subject matter), an area to which they have a closer-range symbiotic relationship; and those (let us say state heritage officials) whose role is necessarily more intellectual, more detached and developed with maturation. In recognizing this distinction between modes of perception it is easy to begin to understand the tensions between national overview and strategic priorities on the one hand, and 'nimbysim'[23] on the other – 'nimbyism', incidentally, that is closely tied to Antony Giddens notion of ontological security, that a certain level of familiarity and routine engenders confidence. As Grenville points out, the familiar and cherished scene is important not just or perhaps even mainly, for its aesthetic and historic significance but for its contribution to everyday identity.[24]

Becoming Autocentric

As we have seen, in England the regulatory and administrative systems by which heritage is managed are currently under review. Heritage protection is being reviewed by the Government department responsible, taking advice from its statutory advisors at English Heritage and the wider heritage community; and planning too is being reformed, with that process becoming more participatory in nature; more democratic. But how can the move towards a symmetrical approach

23   Nimby – 'Not in my Back-yard'.
24   Jane Grenville, 'Conservation as psychology: ontological security and the built environment', *International Journal of Heritage Studies* 13, 6 (2007): 455. See also Peter Read, *Returning to Nothing* (Cambridge: Cambridge University Press, 1996).

to heritage management be promulgated even more effectively? What has worked so far, and what could still be improved?

Research agenda determine and document research priorities for particular geographical areas or subjects. These are often written by experts and serve as a basis for decision making – what are the research priorities; where are competing resources best directed? A symmetrical approach would ensure these agenda take account of multiple views, and promote diverse and multidisciplinary approaches and innovative methodologies. The process of producing agenda would be an inclusive and democratic process, hearing all views and reaching a consensus on priorities. Currently a partnership involving English Heritage, the universities of Bristol and University College London, and Atkins Heritage is co-ordinating an agenda for the later twentieth-century. A website exists[25] and a document was produced which asks:

- What do you remember most clearly about the twentieth century? How are those events or activities still represented in the landscape?
- What do you appreciate, dislike or miss about the later twentieth-century landscape?
- What do you think about change and creation? Would you prefer our landscape to be more like it was in the early twentieth century?
- What can and what should we do with modern landscape character? What should we be recording now for the future?
- Do you have ideas for engaging your community, school or local society with aspects of the twentieth-century landscape?

This consultation was the first stage in a process leading towards a research agenda for the archaeology of the contemporary past. As a process it was comprehensive, democratic and participatory. Crucially the programme recognizes landscape in these democratic terms: as an area, whether rural or urban, ordinary or degraded, special or mundane, as perceived by people and whose character is the result of the action and interaction of natural and/or human factors.[26] 'Change and Creation' has been summarized by Fairclough.[27] He describes it as a programme that considers landscape change as an act of creation as well as of loss, and that recent, current and future changes need to be understood in their own right and – insofar as it is possible – on their own terms. 'Contemporary archaeology' is taken to be not a date range or a period but a state of mind, intimately linked to how people live in the world (that is, to landscape), and to the resource management of all periods of

---

25   www.changeandcreation.org.

26   Council of Europe 2000, article 1(a) of the *European Landscape Convention.*

27   Graham Fairclough, 'The contemporary and future landscape: change and creation in the later twentieth century', in *Contemporary and Historical Archaeology in Theory: Papers from the 2003 and 2004 CHAT Conferences,* edited by L. McAtackney, M. Palus and A. Piccini (Oxford: Archaeopress, 2007), 83.

the past as they contribute to the here-and-now. Thus, as Graves Brown points out in his example of an out-of-town retail park in south Wales, historic landscapes in the Romantic tradition can sit comfortably alongside liminal and interstitial landscapes, places on the periphery with shades of exclusion and transgressive acts, where people shape the landscape by creating paths across neatly mowed lawns and flower beds for example.[28] In both cases landscapes change, there are elements of privatization and control, and the ephemeral is plainly in evidence. Both types of landscape can be valued, by different groups of people and for very different reasons. And this is the key point. Who is to say that the landscape of a retail park in south Wales is any more or less valuable than a landscape, say, characterized by prehistoric ritual monuments, or fields and settlement (Figure 5.3). The official view might be obvious, but is it right? The local view may be very different. More local residents may express support for the former than for the latter, a reflection in part of higher population density in urban and peri-urban areas, and people's lack of engagement with, and unease when faced with 'countryside'.

So here, in a nutshell, is the dilemma and the challenge of finding symmetry in heritage management practices, and the difficulty with being autocentric: the key in a symmetrical (and autocentric) approach to management is to balance these views and findings, or create a system where both can comfortably co-exist. My example is extreme and contentious, and perhaps deliberately chosen to be so, but it highlights the fundamental fiction of much contemporary heritage management practice: that the past is past, and that monuments, places and landscape exist only in the sense of the special, the exceptional, the interesting (to whom?) and the aesthetically pleasing.

But thankfully, in the UK at least, the identification of heritage resources has started moving beyond the historic, and beyond the recognition only of specific culturally-valued heritage 'assets'. We do of course recognize the conventional historic resources – buildings and monuments that have legal protection by virtue of their 'national importance' – but we also now recognize the full range of components that make up the historic environment: elements of landscape character for example, intangible heritage (musical traditions, cuisine, dress and fashion, dialect) and modern heritage.

The criteria for assessing heritage places traditionally give recognition to those which survive best, those with archaeological potential, with historical associations, those which are best documented etc. But increasingly social significance is being drawn into the equation. This is especially the case in Australia where the New South Wales National Parks and Wildlife Service has explored this more than most. Research has examined the ambiguous relations between recent migrant groups (the Macedonian community, the Vietnamese) and the national parks, in an attempt to improve management of the parks and promote people's enjoyment of

---

28   Paul Graves Brown, 'Concrete islands' in *Contemporary and Historical Archaeology in Theory: Papers from the 2003 and 2004 CHAT Conferences*, edited by L. McAtackney, M. Palus and A. Piccini (Oxford: Archaeopress, 2007), 75–82.

Figure 5.3   Symmetry (3). How and in what terms do any of us value the world around us, the landscapes we encounter whether in our professional or personal lives? And why should one person's views hold greater sway than anyone else's? Other than by consensus and through dialogue and negotiation, why should any one group claim intellectual superiority over another in deciding the future of a place that matters? Which of these two landscapes do you feel closest to, and (a very different question) which (if either) would you rather keep – Down Tor (Dartmoor), with cairns and stone row, or a retail park at Trostre, south Wales?

Source: © Dave Hooley (Dartmoor); Paul Graves Brown (Trostre)

them.[29] Underlying this is the key principle of heritage as social action: as James Clifford has said, 'twentieth-century identities no longer presuppose continuous cultures or traditions. Everywhere individuals and groups improvise local performances from (re)collected pasts, drawing on foreign media, symbols and languages'.[30] Cultures are inventive in other words and heritage officials and agencies need to keep abreast of that invention. There is a tendency to think of the heritage as something we protect, keep an eye on, and restore. What Clifford refers to is the way we use heritage as a resource in the ongoing project of creating identity. Denis Byrne describes this as heritage being *deployed* in this aspect of social life.[31]

We can see this to some extent in a recently completed study of wall art recorded on abandoned military bases in England.[32] English Heritage and others have recorded British wall art, alongside that of German and Italian Prisoners of War, Soviet conscripts in eastern Germany, and American servicemen stationed in England during the later years of the Cold War. The American wall art often has a strong Hispanic influence, and can closely resemble the street-gangsta style graffiti of inner cities like Los Angeles from which many servicemen were recruited post-Vietnam. Here, in these modern wall paintings on typically extensive and abandoned 'brownfield' sites now ripe for redevelopment, is a wealth of data about society, cultural diversity and identity within the particular, peculiar and fascinating world of the militarized landscape.

There are multiple ways of protecting the sites that merit it, and for which a form of protection is appropriate or necessary – some are statutory; others not. There are incentive schemes for example in the UK, encouraging farmers to maintain historic character or preserve field monuments. Education ultimately gives owners a pride in what they own and administer, and communities an appreciation of value and meaning. By understanding and valuing things, people are better able to manage them in an appropriate way. Thus protection – in theory at least – doesn't have to be state-led, authoritarian or bureaucratic. Locally and community driven initiatives could be just as effective.

Ultimately historic sites and landscape can reveal and illustrate stories about their past. But these should not be confined to a single official view of the past. Rather they should contribute to multiple narratives or what has been termed alternative histories or archaeology. As Roger Thomas has argued recently,[33] the

---

29   For example, Martin Thomas, *A Multicultural Landscape: National Parks and the Macedonian Experience* (Sydney: NSW National Parks and Wildlife Service, 2001).

30   James Clifford, *The Predicament of Culture* (Cambridge [Mass.]: University of Harvard Press, 1988), 14.

31   Denis Byrne, 'Heritage as social action', in *The Heritage Reader* (op cit.)

32   Wayne D. Cocroft et al., *War Art: Murals and Graffiti – Limitary Life, Power And Subversion* (York: Council for British Archaeology, 2006).

33   Roger M. Thomas, 'Archaeology and authority in the twenty-first century', in *The Heritage Reader*, edited by Graham Fairclough, Rodney Harrison, John H. Jameson Jnr., and John Schofield, (London: Routledge, 2008), 139–148.

extension of choice has affected what people consume (in education, health, and leisure for example) and what they think. A greater plurality exists now in ideas, interests and belief systems than before. In the UK at least, interest in the past has never been greater – so it is no surprise therefore that people think about it in many different ways. A symmetrical archaeology recognizes and encourages multiple narratives.

Capturing Perceptions

Whether we like it or not we live in a world where most people own a mobile phone, and where many of us experience places in the world virtually, through *Google Earth* and similar. We also seem increasingly fascinated with our personal pasts, and the impact (social and physical) that we have made on the world. Hence the success of social networking sites that reunite school friends and work colleagues, and sites that represent these connections in place: where we met; our first date etc. Rather than attempt a comprehensive review of these initiatives and technological developments (which would in any case be out-of-date by the time of publication, this being a subject rewritten by the day) a short review of three projects will suffice.[34]

First is the popular *Map My London* website,[35] part sponsored by the Museum of London and which invites people to submit text, photos, sounds and video on the themes of: Love and Loss, Beauty and Horror, Friendship and Loneliness, Joy and Struggle, and Fate and Coincidence. By adding information to specific places (street corners, schools, pubs etc) through attachments uploaded from mobile phones, a memory map is created whose interest exists at two levels: the points of detail, of interaction and memory at specific points on the map; and the accumulated memory map, of collective memory amongst its (presumably mostly younger) participants. The point is that such maps are easy to establish and maintain, and the information uploaded is easily accumulated and made available for scrutiny. It is completely unscientific and raw; but it is personal, and autocentric, and far from any officially sanctioned view of London's past, and what amongst its streets and buildings holds value to society.

A more systematic and scientific approach is Christian Nold's bio- or emotion-mapping.[36] Nold explains the approach thus: that common everyday maps typically show static architecture and exclude people who inhabit and create the place. His *San Francisco Emotion Map* attempts to remedy this by mapping the space of human perception and experience. In all, 98 participants took part, walking around

---

34   See also John Schofield, 'The new English landscape: everyday archaeology and the Angel of History', *Landscapes* 8, 2 (2007): 115–117.

35   www.mapmylondon.com.

36   www.biomapping.net. See also Nold's *San Francisco Emotion Map* (2007): www. sf.biomapping.net.

the area using Nold's custom-built Bio Mapping device, which documents the wearer's Galvanic Skin Response, an index of emotional response, with a GPS, thus mapping a subtle understanding of the wearer's experience. As with *Map my London*, the results can be interpreted at an individual level, but arguably become more meaningful when accumulated. When looking at the entire map, for example, there appears a general arousal gradient from high in the city centre to low near the edges. Areas of red show hotspots of communal arousal, while darker dots show areas of communal calm. A vivid red cluster can be seen at the intersection of 24th and Mission, for example. This intersection, centred around a major train station, is busy with social interactions. Here the participants often spoke of evangelists, skateboarders and demonstrators as well as the people outside the local McDonalds. There are also clusters around three parks in the area, places that provide inspiring views of the city as well as provoking reflection on past memories.

Finally is 'Snout', a collaboration between the Institute of International Visual Arts, Proboscis and researchers from Birkbeck College London, which explored relationships between the body, community and the environment,[37] building on a previous collaboration on Feral Robots (with Natalie Jeremijenko[38]), to investigate how data can be collected from environmental sensors as part of popular social and cultural activities.

The website describes that by scavenging free online mapping and sharing technologies as a form of 'guerilla public authoring', the project explores how communities can gather and visualize evidence about local environmental conditions and how that information can be used to participate in or initiate local action. Snout created two prototype sensor *wearables* based on traditional carnival costumes. Carnival is a time of suspension of the normal activities of everyday life – a time when the fool becomes king for a day, when social hierarchies are inverted, a time when everyone is equal. There is no audience at a carnival, only carnival-goers. Snout involved 'participatory sensing' as a lively addition to the popular artform of carnival costume design, engaging the community in an investigation of its own environment, something usually done or organized only by local authorities and state agencies.

Everyday Heritage

My contention here is that a symmetrical approach provides a useful context within which to question some of the more fundamental and established principles of heritage management practice – why the heritage needs to be managed at all, and for whose benefit; why some heritage places need protection; and questioning the idea of the past as a renewable resource; and the past, present and future as a continuous process. Increasingly those employed or engaged in heritage

---

37   http://socialtapestries.net/snout/.
38   http://socialtapestries.net/feralrobots/index.html.

management practice see the past as something we (society) actively engage with, not something only a select group of heritage managers pontificate upon and 'manage'. That's what the heritage should be about: the everyday, the everywhere and something for (and of) everybody. The time for elite heritage has long passed, if it ever really existed at all.

Acknowledgements

I owe thanks to a number of friends and colleagues who have helped me to shape some of the ideas expressed in this chapter. Notable of these are members of English Heritage's Characterization Team who in our meetings and discussions over the years have – perhaps unwittingly – encouraged the views of landscape and heritage that I describe here; and those in the Change and Creation group, all of whom are nowhere near the box, let alone living and thinking outside it. Graham Fairclough is in both of these groups, so I'm doubly indebted to him. I would also like to thank Christian Nold and Giles Lane for discussing their work with me and exploring opportunities for collaboration with such enthusiasm. This is a suitable opportunity also to thank undergraduate and postgraduate students at the universities of Southampton, Bristol, Flinders, Turku and Oulu who over the past two to three years have listened to these views with patience and good grace, and commented intelligently, persuading me to think further. It is a process that will, I hope, continue. Images were supplied by David Green, Paul Graves Brown and Dave Hooley.

Bibliography

Byrne, Denis. 'Heritage as social action' in *The Heritage Reader*, edited by Graham Fairclough, Rodney Harrison, John H. Jameson Jnr., and John Schofield, 149–174. London: Routledge, 2008.

Clifford, James. *The Predicament of Culture*. Cambridge (Mass.): University of Harvard Press, 1988.

Cocroft, Wayne D., Danielle Devlin, John Schofield, and Roger J.C. Thomas. *War Art: Murals and Graffiti – Military Life, Power And Subversion*. York: Council for British Archaeology, 2006.

Council of Europe, *European Landscape Convention*, 2000, http://conventions. coe.int/Treaty/en/Treaties/Html/176.htm.

Dubrow, Gail. 'Feminism and multicultural perspectives on preservation planning', in *Making the Invisible Visible: A Multicultural Planning History*, edited by L. Sandercock, 57–77. Berkeley: University of California Press, 1998.

Fairclough, Graham. 'The contemporary and future landscape: Change and creation in the later twentieth century', in *Contemporary and Historical Archaeology in Theory: Papers from the 2003 and 2004 CHAT Conferences*, edited by L. McAtackney, M. Palus and A. Piccini, 83–88. Oxford: Archaeopress, 2007.

Gard'ner, James M. 'Heritage protection and social inclusion: A case study from the Bangladeshi community of East London', *International Journal of Heritage Studies* 10, 1 (2004): 75–92.

Graves Brown, Paul. 'Concrete Islands', in *Contemporary and Historical Archaeology in Theory: Papers from the 2003 and 2004 CHAT Conferences*, edited by L. McAtackney, M. Palus and A. Piccini, 75–82. Oxford: Archaeopress, 2007.

Grenville, Jane. 'Conservation as psychology: Ontological security and the built environment', *International Journal of Heritage Studies* 13, 6 (2007): 447–61.

Hannerz, Ulf. *Soulside: Inquiries into Ghetto Culture and Community*. Stockholm: Almqvist and Wiksell, 1969.

Harrison, Rodney. *Shared Landscapes: Archaeologies of Attachment and the Pastoral Industry in New South Wales*. UNSW Press: Department of Environment and Conservation (NSW), 2004.

Hayden, Dolores. *The Power of Place: Urban Landscapes as Public History*. Cambridge (Mass.): The MIT Press, 1995.

Porteous, J. Douglas. *Environmental Aesthetics: Ideas, Politics and Planning*. London and New York: Routledge, 1996.

Read, Peter. *Returning to Nothing*. Cambridge: Cambridge University Press, 1996.

Schofield, John. 'Intimate engagements: Art, heritage and experience at the "place ballet"', *International Journal of the Arts in Society* 1, 5 (2007a): 105–114.

Schofield, John. 'The New English Landscape: Everyday Archaeology and the Angel of History', *Landscapes* 8, 2 (2007b): 106–125.

Schilling, Kath. *Aboriginal Women's Heritage: Nambucca*. NSW National Parks and Wildlife Service, 2003.

Thomas, Martin. *A Multicultural Landscape: National Parks and The Macedonian Experience*. NSW National Parks and Wildlife Service, 2001.

Thomas, Roger M. 'Archaeology and authority in the twenty-first century', in *The Heritage Reader*, edited by Graham Fairclough, Rodney Harrison, John H. Jameson Jnr., and John Schofield, 139–148. London: Routledge, 2008.

Tuan, Yi-Fu. *Space and Place: The Perspective of Experience*. Minneapolis: University of Minnesota Press, 1977.

Webmoor, Tim. 'What about "one more turn after the social" in archaeological reasoning? Taking things seriously', *World Archaeology* 39, 4 (2007): 563–78.

Witmore, Chris L. 'Symmetrical archaeology: excerpts of a manifesto', *World Archaeology* 39, 4 (2007): 546–62.

Chapter 6

Reputation and Regeneration: History and the Heritage of the Recent Past in the Re-Making of Blackpool

John K. Walton and Jason Wood

British seaside resorts present particularly difficult problems as regards the 'heritage of the recent past'. In the first place, their experience and circumstances raise contentious questions under the headings both of 'recent' and of 'heritage'. Secondly, their current economic difficulties, and problematic relationship with central Government, add further dimensions to the question of what is to be done to regenerate these places, which are peripheral by definition but owe their existence and identity as resort destinations and residential areas to external demand for what they have to offer, if not from the metropolis then from regional cities and their satellite towns. Thirdly, proposals for regeneration are sometimes threatening in themselves to the identities of such towns as seaside resorts, to components of their 'heritage' in that guise, and especially to those aspects of the 'heritage of the recent past' that are not obviously iconic in the manner of Bexhill's De La Warr Pavilion or Morecambe's Midland Hotel.[1]

The popular Lancashire resort of Blackpool is of especial interest in all these respects, as its unique history as the world's first working-class seaside resort has left a sufficient legacy of seaside architecture, public spaces and entertainment traditions to merit serious consideration as a potential World Heritage Site under UNESCO's 'cultural landscape' rubric (Figure 6.1). But such a proposal, despite the flexibility of the category, generates possible tensions with proposals for regeneration through modernization under the auspices of the urban regeneration company ReBlackpool and the local authority's Masterplan. These questions have arisen, and become pressing, at a time when the resort's holiday and leisure industries appear to be suffering from a deepening crisis. Its symptoms include sharply falling visitor numbers, average spends and length of stay, sustained decay in aspects of the public realm (despite recent successful investment in significant aspects of the resort environment), persistent and often ill-informed or uninformed hostility from sections of the media, and mixed messages from central Government

---

1 A. Fairley, *De La Warr Pavilion: the Modernist Masterpiece* (London: Merrell, 2006); B. Guise and P. Brook, *The Midland Hotel: Morecambe's White Hope* (Lancaster: Palatine Books, 2007).

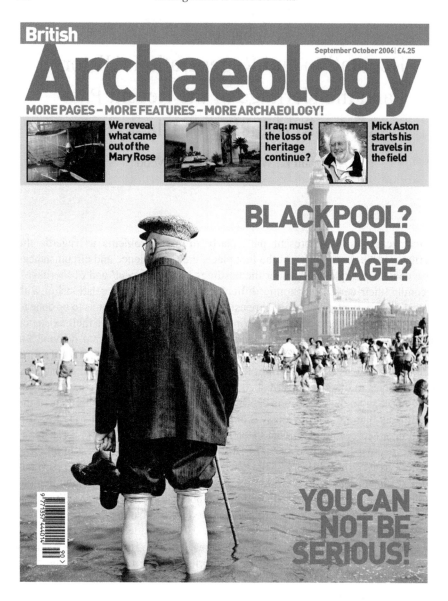

Figure 6.1   The front cover of *British Archaeology* published in 2006 featuring our article on Blackpool's ambition for World Heritage Site status
Source: © *British Archaeology*

on policy initiatives whose outcomes have been mainly negative. Most obvious among these was the rejection in 2007 of the town's bid to be nominated as the site of the UK's first so-called 'super-casino'.

This chapter looks at Blackpool's efforts to regenerate itself, combining innovation and the search for new markets with an appeal to tradition, identity and the 'heritage of the recent past', as it begins to implement a heritage strategy, seeks inscription as a World Heritage Site, trades on the industrial archaeology of the holiday industry and tries to generate alternative income streams for regeneration following the decision not to grant the casino licence.[2] It also investigates conflicting media attitudes to these strategies and to the changing nature of Blackpool itself as a seaside resort at the beginning of the new millennium.[3]

The 'Heritage of the Recent Past'?

The concept of the 'heritage of the recent past' in this setting requires some introductory and contextual discussion. Definitions of the 'recent past' are likely to be arbitrary, but a sensible cut-off date for present purposes is probably the end of World War I (i.e. just within living memory), although the growing vogue for some kinds of inter-war building (especially lidos and anything that can be labelled 'Modernist' or 'Art Deco') might be identified with a growing cultural acceptability that betokens the development of a historical or nostalgic perspective and undermines the notion of 'the recent'.[4] As in other British seaside resorts, much of Blackpool's surviving architectural heritage is Victorian in origins, and thus might not fall even within this relatively generous canon; but the full picture is much more complicated. Among the important surviving entertainment centres, the Winter Gardens dates back to the mid-1870s, but almost all of it is a subsequent addition or has been rebuilt or extensively altered, sometimes more than once. The present Opera House, completed in 1939, is the third one at the Winter Gardens; the Pavilion Theatre, originally opened in 1878, has been through several incarnations including adaptation as a cinema in the 1930s, while in 1986 its stage became the Palm Court Restaurant; the Indian Lounge of 1896 became the Planet Room in 1964 and then the Arena in the 1980s; the Empress Ballroom was modified in the 1970s in response to falling demand for formal dancing; the

2   J. K. Walton and J. Wood, 'World heritage seaside', *British Archaeology* 90 (2006), pp. 10–15; J. Wood and J.K. Walton, 'La Station Balnéaire Comme Site du Patrimoine Mondial? Le Cas de Blackpool', in Y. Perret-Gentil, A. Lottin and J.-P. Poussou (eds), *Les Villes Balnéaires d'Europe Occidentale du XVIIIe Siècle à Nos Jours* (Paris: Presses de l'Université Paris-Sorbonne, 2008), pp. 423–52.

3   J.K. Walton and J. Wood, 'History, heritage and regeneration of the recent past: The British context', in N. Silberman and C. Liuzza (eds), *Interpreting the Past V, Part 1: The Future of Heritage: Changing Visions, Attitudes and Contexts in the 21st Century* (Brussels: Province of East-Flanders, Flemish Heritage Institute and Ename Center for Public Archaeology and Heritage Presentation, 2007), pp. 99–110.

4   J. Smith, *Liquid Assets: The Lidos and Swimming Pools of Britain* (London: English Heritage, 2005); F. Gray, *Designing the Seaside: Architecture, Society and Nature* (London: Reaktion, 2006).

Figure 6.2   The Winter Gardens entertainment complex, originally by Thomas Mitchell, incorporates elements built variously between 1875 and 1939. The Spanish Hall of 1931 is lavishly decorated in fibrous plaster by Andrew Mazzei, Art Director of the Gaumont Film Company

Source: © Jason Wood

Galleon Bar, the Baronial Hall and the Spanish Hall are survivors from 1931 (Figure 6.2). This partial potted history of a complex site shows how the architecture and decoration of the 'recent past' can be serially overlaid on earlier features, in ways that are particularly characteristic of the entertainment industry, with its urgent commercial imperative to keep abreast of or ahead of changing fashions and preferences in popular taste.[5] The three piers, dating from 1863, 1868 and 1893, have all experienced recurrent changes to their attractions and superstructure. The North Pier lost its ornate Indian Pavilion of 1874 to a fire in 1919, and its successor went the same way in 1938. The Central and South (originally Victoria) Piers have added fairground rides as well as regularly updating leisure buildings on the superstructure.[6] The resort's signature structure, the Tower of 1894, has itself undergone many changes in the internal arrangements and functions of the large

---

5   http://www.blackpool.gov.uk/services/M-R/PlanningConservationAreasandListed Buildings accessed 20 April 2008.

6   http://www.theheritagetrail.co.uk/piers/blackpool accessed 20 April 2008.

Figure 6.3   A typical street of boarding-houses. Blackpool reputedly still has more beds for hire than the whole of Portugal
Source: © Jason Wood

brick building in which the entertainments are housed, including the abandonment of the menagerie and of live animal acts in the Circus; and the ornate Ballroom was completely reconstructed, including the decorative plasterwork, after a serious fire in 1956.[7]

The 'recent past', perhaps especially under the circumstances of the seaside entertainment industry, is in practice a movable feast; and even more so if we include Blackpool's industrial landscape of several thousand boarding-houses for seaside visitors (Figure 6.3), most of them purpose-built in the late nineteenth century, some in the 1930s, and all incorporating complex sequences of refurbishment as extra bedrooms were added in the eaves and at the back, updated plumbing entailed tangled lattices of additional pipes, and internal remodelling responded to changing visitor expectations about privacy and en-suite bath and shower rooms.[8] Most of Blackpool's buildings are a legacy of continuous embellishment,

---

7   http://www.theblackpooltower.co.uk/facts.htm accessed 20 April 2008.

8   J.K. Walton, 'The Blackpool landlady revisited', *Manchester Region History Review* 8 (1994), pp. 23–31.

improvement, piecemeal replacement, and responses to damage and disaster over a long period, with only the integrity of the core structures (themselves subject to rolling replacement) sustaining 'authenticity' in a literal-minded sense; and such changes have usually arisen from perceived commercial necessity, as part of the shifting culture of entertainment or accommodation provision in a popular resort, which is necessarily resistant to the preservation of physical 'heritage' as a dominant value in itself, for its own sake or even (in most cases) as part of the enterprise's own appeal.

Questions of Legitimacy

The heritage of the seaside holiday industry also carries a significant perceived legitimacy deficit, especially when the forms in which it manifests are low-key and easily disparaged. A good example is the car park and bus station in Blackpool's Talbot Road, built in 1937–39 and described by the English Heritage seaside historians as 'a rare example of a multi-storey car park outside London or a major industrial city before the Second World War'.[9] It does not seem to occur to them that at this point Blackpool was itself a 'major industrial city', whose dominant economic activities, on a unique scale, were tourism, leisure and accommodation. The building itself, neglected for years and slated for demolition in the current re-making of Blackpool's 'Talbot Gateway', was never a serious candidate for listing, historically significant though it was. Industrial archaeology as more commonly understood, that of the 'Industrial Revolution' of the eighteenth and nineteenth centuries, of manufacturing and transport, ironworks, steam-powered factories, canals and railways, has made headway in overcoming these problems since the 1960s, when the pioneer Beamish open-air industrial museum attracted denigration and derision for harking back to a dirty, undignified, painful, undesirable past and desecrating a rural area in the process.[10] But these attitudes die slowly, as witness Bradford's recurring problems with the derisive triviality of the media as it attempted to reinvent itself as an industrial heritage tourism destination or the enduring media assumption that legitimate aspirants to World Heritage Site status are limited to such icons as the Taj Mahal and the Great Wall of China, while industrial inscriptions such as Blaenavon are a matter for surprise and mockery.[11]

---

9   A. Brodie and G. Winter, *England's Seaside Resorts* (Swindon: English Heritage, 2007), p. 61.

10   G. Cross and J.K. Walton, *The Playful Crowd: Pleasure Places in the Twentieth Century* (New York: Columbia University Press, 2005), chapter 6.

11   D. Russell, 'Selling Bradford: Tourism and northern image in the late twentieth century', *Contemporary British History* 17 (2003), pp. 49–68; S. Morris, 'Forget the Taj Mahal and Pyramids: Blaenavon is the Place to Visit', *The Guardian*, 26 March 2008, p. 9.

The frontier of legitimacy moves at different speeds in different settings and for different aspects of surviving 'heritage', even at the seaside. The significance of certain kinds of seaside heritage was recognized in other resort settings long before it became an issue at Blackpool. The humblest cottages in the imagined 'fishing quarters' of Whitby and St Ives already had articulate and well-organized defenders against proposals for demolition and redevelopment in the 1930s,[12] while Brighton's Regency Society was founded as early as 1945 to protect the resort's Georgian terraces against redevelopment for flats, followed in 1960 by the Hove Civic Society which organized local residents against similar proposed demolitions. By the mid-1970s residents of Brighton's North Laine area, a distinctive group of early Victorian streets which was threatened by wholesale demolition, were banding together in search of Conservation Area status, and starting their own newsletter as a focus for their successful campaign.[13]

But these were campaigns that gained legitimacy from the historical patina, if not always the architectural distinction, of the places they sought to conserve. Blackpool's holiday and entertainment industry suffered from a shortage of credibility on both counts, as well as a paucity of campaigners. This helps to explain the loss of several iconic entertainment buildings during the last quarter of the twentieth century, as we shall see. It was not until 1975 that the town acquired a Civic Society, and it was never a powerful local organization when set against the local authority and the entertainment industry. Earlier campaigns against a post-war town centre redevelopment plan had been mainly on the grounds of disruption to existing businesses, and it was the mobilization in 1973 of the 'Friends of the Grand Theatre' to save this Frank Matcham masterpiece from demolition and restore it to live theatre use that marked the most plausible emergence of sustained and determined 'heritage' campaigning in Blackpool. Even in this case, the agenda was more theatrical than architectural, although Brian Walker's published opinion in 1980 that the Grand was 'one of the finest that Matcham designed' was music to the ears of the 'Friends'. Moreover, the campaigners' attention to the continuous history of theatrical performance on the site marked out a concern for a lived, and living, version of the 'heritage of the recent past' that was to become significant on a broader front. We shall return to these issues: what matters for the moment is the delayed and enduringly contested nature of perceptions of the cultural legitimacy of seaside entertainment heritage, architectural or otherwise, in the Blackpool setting, even as compared with other seaside resorts.[14]

---

12   J. K. Walton, *Tourism, Fishing and Redevelopment: Post-war Whitby, 1945–1970* (University of Cambridge: Institute of Continuing Education, Occasional Paper No. 5, 2005), pp. 5–31.

13   http://www.regencysociety.org accessed 25 April 2008; http://www.hovecivic society.org accessed 25 April 2008; http://www.nlcaonline.org.uk/html/history_rewriting. html accessed 25 April 2008.

14   http://www.blackpoolgrand.co.uk/information/6/273/How-the-Grand-was-Saved. htm accessed 26 April 2008, reminiscences of Burt Briggs; B. Walker (ed.), *Frank Matcham,*

Economic Difficulties and Problems of Government

Questions of seaside regeneration through the 'heritage of the recent past' must also be seen in the context of the economic difficulties faced by many coastal locations, especially provincial popular resorts, in the late twentieth and early twenty-first centuries, and the reluctance of Government to address the issues of falling demand, unemployment, ageing populations and deterioration in the public realm. The well-known economic and demographic analysis by Beatty and Fothergill, issued in 2003, has challenged the more alarmist assumptions about the 'decline of the seaside' between 1971 and 2001 by demonstrating the resilience of many British coastal resorts, and their capacity to attract new migrants and generate new jobs, in tourism as well as other sectors.[15] But there is no doubt that some resorts, or districts within resorts, faced – and continue to face – serious problems. Blackpool was later than some provincial popular resorts in sliding into difficulty, but by the 1991 census it was already occupying high places in league tables of multiple deprivation, especially in central wards which had been the core of its traditional holiday industry; and by the turn of the millennium the resort was demonstrably in crisis, as visitor numbers, length of stay and per capita spending entered a sustained downturn.[16] Beatty and Fothergill's Blackpool case-study shows that the town had performed less well than most other substantial British seaside resorts, whether the preferred indicators involved unemployment, in-migration or female economic inactivity.[17]

Blackpool's economic performance was thus below the average for British seaside resorts on a range of indicators; but the actual figures were positive and, in themselves, encouraging. The two leading 'headline' conclusions of the research team were that, 'The tourist component of the Blackpool economy should certainly not be written off as a lost cause', and that, 'Efforts to generate new jobs in Blackpool should be strongly endorsed.'[18] If this diagnosis was accepted, the question was how to go about revitalizing Blackpool's tourist economy, and what form the new jobs should take. Blackpool Council had already unveiled a Masterplan, pursued through the urban regeneration company ReBlackpool, initially chaired by the distinguished planning academic Professor Sir Peter Hall.[19] The Masterplan originally envisaged the funding of a thoroughgoing regeneration programme through an income stream

---

Theatre Architect (Belfast: Blackstaff Press, 1980), pp. 125–6; B. Band, *Blackpool Grand Theatre 1930–1994* (Lytham: Barry Band, 1994).

15   C. Beatty and S. Fothergill, *The Seaside Economy* (Centre for Regional Economic and Social Research, Sheffield Hallam University, Sheffield, 2003).

16   J.K. Walton, *Blackpool* (Edinburgh: Edinburgh University Press, 1998).

17   C. Beatty and S. Fothergill, *A Case Study of Blackpool*, Seaside Town Research Project, Paper No. 3 (Centre for Regional Economic and Social Research, Sheffield Hallam University, Sheffield, 2003), pp. 19, 33–5.

18   *Ibid.* p. 5.

19   Blackpool Council, *New Horizons: Blackpool Resort Masterplan* (Blackpool, 2003).

derived principally from casino gambling – a possibility signposted at the time by Government thinking – and promised to re-define Blackpool's future as a top-quality, world-class resort destination. This immediately generated potential problems involving expropriation and site clearance, as well as the inevitable arguments about the form and content of the new buildings and attractions. In the event the Government's decision on the location of the much-anticipated 'super-casino' went against Blackpool in favour of Manchester, but this too has now been abandoned following a reversal of Government policy.[20] However, the decision not to grant Blackpool the casino licence produced an unexpected windfall in the shape of a Blackpool Task Force, led by the North West Regional Development Agency, set up to rethink long-term regeneration plans for the town.[21] The result is a revised Masterplan which still represents the single most comprehensive programme of developments since the park, promenade improvement and planning proposals of the 1920s and 1930s.[22] However, given the lack of central Government support for Blackpool, it is questionable whether the necessary funding to deliver the regeneration plans will be forthcoming.

Heritage, Regeneration, Uncertainty and Conflict

The use of history and 'heritage' in urban regeneration through cultural tourism and associated 'gentrification' poses problems of direct relevance to Blackpool, which have generated an extensive literature, wittily pulled together by Melanie K. Smith in her introduction to a recent collection of essays. She points out that the up-market 'creative' tourists who spend heavily and enhance destination status require 'more authentic indigenous or authentic venues' as opposed to the proliferation of 'international blandscapes' and 'generica', or the 'serial monotony' of 'standardized developments that could be anywhere'. She also offers a series of dichotomies that might form the basis for further elaboration in the light of local circumstances: tensions between the global and the local, between standardization and 'place-making', the preservation of 'heritage' and the promotion of contemporary culture.[23] The 'heritage of the recent past', or perhaps that of the continuing and evolving past, can be called in to reconcile these contradictions, at least in the particular setting of Blackpool.

---

20   Blackpool Council, *Towards a World Class Resort Destination: Submission to the Casino Advisory Panel* (Blackpool, 2006); Casino Advisory Panel, *Final Report of the Casino Advisory Panel* (London, 2007).

21   North West Regional Development Agency, *Blackpool: An Action Plan for Sustainable Growth. Report of the Blackpool Task Force* (Blackpool, 2007).

22   ReBlackpool, *Blackpool Resort Masterplan: Rethink, Rebuild, Renew* (Blackpool, 2007).

23   M.K. Smith (ed.), *Tourism, Culture and Regeneration* (Wallingford: CABI, 2007), pp. xiv–xvii, 4–5.

At the beginning of the twenty-first century Blackpool is passing through a difficult period of declining older markets and problematic new ones, but its present Council is making determined efforts at regeneration in ways that respect not only its traditions, which have a potent market value of their own, but also its 'tradition of invention' and innovation, as expressed enduringly in the town's motto, 'Progress'. In taking the Masterplan forward the Council accepts it needs to strike the right balance; 'it is important to regenerate and innovate to make Blackpool a modern resort but we also recognize the importance of our heritage, which in itself attracts visitors'. The Council also acknowledges that its actions will be under scrutiny as 'the heritage of Blackpool has a place in the whole nation's consciousness.'[24]

Changes to the physical and social face of the town are inevitable so a well-informed appreciation of the heritage context in which the Masterplan will operate will be essential. An important milestone in this respect was the publication in July 2006 of the Council's first heritage strategy with its mission 'to discover, conserve, learn from and celebrate the past in order to inspire a better future for the town and people of Blackpool.'[25] Setting out an agenda for action over the next five years, the document includes a range of ambitious projects and initiatives exploring the overlap between tourism and heritage interests with the aim of enhancing the value of Blackpool as a place for residents and visitors alike. In particular, the strategy seeks to capture and preserve memories, renovate and celebrate outstanding buildings, safeguard and defend a unique social history, and educate and converse with the public in a variety of exciting ways. A positive start has already been made with a number of oral and community history programmes either under way or being planned.[26] In addition, Professor Vanessa Toulmin, Research Director of the National Fairground Archive at Sheffield University, is working with the Council on the 'Admission All Classes' project to bring themed weekends celebrating the history of past entertainment (cabaret, variety, circus, burlesque) to local venues.[27]

Recently, the Council, with the support of English Heritage, has initiated an important piece of research in the form of the Blackpool Historic Townscape Characterization Project. While the major buildings and places of the entertainment and leisure industry are well understood, appreciation of the rest of Blackpool is less well developed, especially for the large boarding-house districts and residential areas. The project sets out to map and characterize a number of key areas of the historic townscape to ensure that their value and significance permeates through to generate effective policies so that planning, development and tourism decisions are based on informed knowledge, understanding and respect for what has gone before

24 Blackpool Council, *Vision for Blackpool: Blackpool's Community Plan 2004–2020* (Blackpool, 2004) p. 2.

25 Blackpool Council, *Heritage Strategy 2006–10* (Blackpool, 2006) p. 14.

26 *Ibid.*, p. 18.

27 http://www.admissionallclasses.com accessed 28 April 2008.

and people's interest in and attachment to it. It is also intended that the character maps will eventually help to define a potential World Heritage Site boundary and buffer zone. At a strategic level the project will make a significant contribution to the further development and implementation of the Masterplan. At an individual scheme level, through the medium of planning and design briefs, the information generated will help influence and encourage imaginative developments that either secure the future reuse or enhancement of heritage assets or inform a new high quality architecture that responds to and respects the resort's distinctive heritage and character.

A pre-emptive example of this kind of collaboration in action was the bidding brief for ReBlackpool's application in 2006–7 for Big Lottery funding for the 'People's Playground', a major promenade regeneration programme, that emphasized the need to take full account of the town's history and heritage in a 'cultural landscape' sense. Although this was ultimately turned down for funding, the process of bid evaluation involved inputs from a historian as well as an artist and a spectrum of expertise in urban planning, engineering and financial management, with emphasis not only on Blackpool's unique architectural heritage but also on the 'intangible heritage' of its entertainment industry and of informal popular pleasures along the promenade and piers, generating a distinctive holiday atmosphere. Subsequent collaboration has borne fruit in the shape of a £4 million grant from the Government's Sea Change programme towards the creation of a 20,000 capacity outdoor performance space to showcase cultural events on the promenade in front of the Tower.[28]

Blackpool's History and the 'Heritage of the Recent Past'

As part of the processes discussed in the previous paragraph, and specifically in relation to Blackpool's proposed bid for World Heritage Site status, a strong case for the distinctiveness, indeed uniqueness, of Blackpool's history on a global stage can be readily constructed, and harnessed to the exploitation of the 'heritage of the recent (and less recent) past'. In the last decades of the nineteenth century Blackpool became the world's first working-class seaside resort, and in the inter-war years it acquired a national visitor catchment area and developed into a unique centre of popular entertainment.[29] This entailed immense investment in infrastructure (almost all by the local authority), transport facilities, entertainment facilities funded by the 'shareholder democracies' of big limited companies (three piers, the Winter Gardens, the Tower and much else besides), and

---

28 http://www.culture.gov.uk/reference_library/media_releases/5353.aspx (accessed 1 August 2008).

29 J.K. Walton, 'The demand for working-class seaside holidays in Victorian England', *Economic History Review* 34 (1981), pp. 249–65; Walton, *Blackpool*.

accommodation (nearly 5000 boarding-houses at the 1921 census).[30] Blackpool's heyday came in the 1950s and early 1960s, although its entertainments failed to move with the times; and it acquired an equivocal reputation, loved by its loyal customers from industrial Britain but sometimes suffering the condescension of politicians and journalists who attended the party and trade union conferences at the resort.[31] It continued to reinvent itself successfully during difficult times in the late twentieth century, but fell into the difficulties that were common across much of the popular British seaside at the turn of the millennium.[32] Despite its recent problems, Blackpool is still by far the most popular British resort, after a career which saw it attracting over 3 million visitors per annum by the 1890s and over 7 million by the 1930s.[33]

As discussed above, Blackpool has retained its three Victorian piers, together with the Tower and the Winter Gardens, with all their subsequent adaptations to changing tastes, preferences and visiting publics since the decline of the manufacturing industries of northern England, the English Midlands and the Clyde Valley, and their associated town holidays, since the 1970s.[34] The Pleasure Beach, an Edwardian amusement park in origins, but the subject of an impressive Modernist makeover by Joseph Emberton in the 1930s, is similarly a palimpsest of popular culture despite the recent loss of well-established rides, with the Sir Hiram Maxim Captive Flying Machine of 1904 and the inter-war Noah's Ark (Figure 6.4) and Big Dipper rubbing shoulders with the Big One roller-coaster of 1994 and subsequent spectacular rides.[35] The Beaux Arts influences on the municipal Stanley Park, which was opened ceremonially in 1926 and is now the subject of regeneration through the Heritage Lottery Fund (HLF), provide another distinctive strand in Blackpool's entertainment-based 'heritage of the recent past', as do the promenades and their associated maritime parks, gardens, and cliff walks, mainly of inter-war vintage. Particularly remarkable, and difficult to deal with for regeneration purposes, are the long parallel streets of Victorian holiday lodging- and boarding-houses, together with the Edwardian private hotels of the North Shore and the 1930s developments south of the Pleasure Beach. These constitute a unique industrial landscape, without parallel anywhere else in the world, which will be challenging to sustain in the light of changing tastes and preferences, as has been demonstrated (for example) by the difficulty of saving Miami's Art Deco South Beach hotel

30   Walton, 'The Blackpool landlady revisited'.

31   Walton, *Blackpool*.

32   J.K. Walton, *The British Seaside: Holidays and Resorts in the Twentieth Century* (Manchester University Press, 2000).

33   J.K. Walton, 'The world's first working-class seaside resort? Blackpool Revisited, 1840–1974', *Transactions of the Lancashire and Cheshire Antiquarian Society* 88 (1992), pp. 1–30.

34   Walton, *Blackpool*.

35   J.K Walton, *Riding on Rainbows: Blackpool Pleasure Beach and its Place in British Popular Culture* (St Albans: Skelter Publishing, 2007).

Figure 6.4   **The Pleasure Beach's Noah's Ark, opened in 1922 and re-vamped in 1934, was designed by the American William Homer Strickler. It is one of only two surviving in the world. Beyond is Sir Hiram Maxim's Captive Flying Machine of 1904**

Source: © Jason Wood

area.[36] Similarly, the survival of the promenade tramway, unique in England, generates tensions between its role as heritage transport attraction and the need to provide a modernized passenger service for visitors and residents alike (Figure 6.5). These are complex issues, exacerbated by the unique and compelling nature of what has survived in an industry whose buildings and practices are normally, as at Coney Island and Atlantic City, ephemeral. And there is the further problem of deciding at what point really recent additions to the entertainment and hospitality menu, such as the Sandcastle indoor pool complex, the Hilton Hotel or the Pleasure Beach's Big Blue Hotel become the 'heritage of the recent past', alongside the recent public art of the renovated south promenade or the southern road approaches.[37]

Blackpool's distinctive, indeed unique history has not been reflected in the listing of buildings officially perceived to be of architectural or historical importance. As

36   M.B. Stofik, *Saving South Beach* Gainesville: University Press of Florida, 2005).

37   Cross and Walton, *The Playful Crowd*; B. Simon, *Boardwalk of Dreams* (New York: Oxford University Press, 2004).

**Figure 6.5   The pioneer public electric street tramway opened in 1885 and
was municipalized seven years later**
Source: © Blackpool Council

of April 2008, the Tower is Blackpool's only Grade I listed structure. There are
four listings at II*: a Victorian Roman Catholic church, a war memorial shrine,
the Grand Theatre and most of the structures that make up the Winter Gardens.
Of the 28 listings at the lower Grade II, seven are directly connected with the
holiday and entertainment industries. They include the North Pier of 1863 (but
neither of the other piers); the 'New Clifton Hotel' on the site of the early Clifton
Arms; the Victorian and Edwardian Imperial Hotel at North Shore; the Raikes Hall
Hotel, a Georgian mansion turned hotel at the core of the Royal Palace Gardens,
an outdoor entertainment complex that flourished from the early 1870s to the turn
of the century; the former King Edward cinema of 1913; the splendidly florid
former Miners' Convalescent Home of 1925–7; the former Odeon Cinema of
1938–9, a very large and impressive example of its kind whose auditorium was
divided in 1975 to create two smaller cinemas, before conversion at the turn of the
millennium into the Funny Girls cabaret venue; and the White Tower of 1939, the
second Casino building of Blackpool Pleasure Beach, which has twice undergone
extensive renovation and adaptation. The Pleasure Beach, although willing to
market itself as the home of historic as well as futuristic amusement park rides, has

resisted any suggestion of listing any other of its appurtenances, preferring to label its vintage attractions with commemorative and informative plaques based on and using the United States 'National Historical Marker' terminology, which has no official standing in Britain.[38] Blackpool's listed buildings also include two groups of distinctive shelters on the promenade, probably dating from the completion of a major municipal scheme for widening and embellishment in 1905. Of this limited roll-call (alongside seven religious buildings and associated structures, two vernacular cottages and a farmhouse, a windmill and several municipal or other public buildings), only the Odeon and the White Tower, together with the more recent additions to the Winter Gardens, Tower and North Pier, and the former Miners' Convalescent Home, really qualify for consideration under a post-First World War 'heritage of the recent past' rubric. This constitutes a reminder that Blackpool's official interest in its seaside heritage is not of long standing.[39]

Moreover, Blackpool has only two Conservation Areas. Both were designated in 1984, and the one covering the Talbot Square area was renamed the Town Centre Conservation Area when it was extended during 2005–6. The Council's descriptive leaflet represents it as Blackpool's 'civic heart', including as it does the Town Hall and the former central Post Office building as well as the North Pier and the Metropole Hotel, on the site of Bailey's which was one of the first Blackpool buildings to be purpose-built to accommodate visitors. It also extends to the Winter Gardens. It is still limited in size, but it does contain short but significant stretches of characteristic mid-Victorian bay-windowed boarding-houses now converted to shop, office and restaurant use as well as its concentration of assorted listed buildings.[40] The second designated area contains Stanley Park, a Grade II* listed park since 1995, and its surrounding planned boulevard system.[41] Here the 'heritage of the recent past' does come into play: the park's opening in 1926 is still just within living memory, and its fiercely supportive group of 'Friends of Stanley Park' have played an important part in securing HLF support for extensive refurbishment, with Stage 1 approval in 2003 and a new Visitor Centre opening in 2005.[42]

Even this constitutes a significant advance on 1969, when the Blackpool entry in the North Lancashire volume of Sir Nikolaus Pevsner's *Buildings of England* series was shorter than those for Blackburn and Burnley, industrial towns of comparable size and vintage.[43] At this time, as Pevsner noted, 'The Ministry of Housing and

---

38   Walton, *Riding on Rainbows*.

39   http://www.blackpool.gov.uk/services/M-R/PlanningConservationAreasand ListedBuildings (accessed 20 April 2008).

40   http://www.blackpool.gov.uk/services/M-R/PlanningConservationAreasand ListedBuildings (accessed 25 April 2008).

41   H. Meller, *European Cities 1890–1930s* (Chichester: John Wiley and Sons, 2001), pp. 197–202.

42   http://www.friendsofstanleypark.org.uk (accessed 25 April 2008).

43   N. Pevsner, *The Buildings of England: North Lancashire* (Harmondsworth: Penguin, 1969), pp. 61–8, 68–72, 79–84.

Local Government (had) issued no list for Blackpool, which means that there are no buildings of architectural or historic interest in the town. It depends of course on what one means by historical and by architectural... English social history of the second half of the C19 and the first half of the C20 cannot be written without Blackpool.' The first listings came soon afterwards, just in time to help the 'Friends of the Grand' to organize their successful resistance to the proposed demolition of the theatre. Pevsner's observation about social history was prescient, but '*The Buildings of England* is not about social historiography', and Pevsner did not find Blackpool to be either attractive or interesting. His brief comments on individual buildings are often damning: 'the lack of aesthetic discrimination which we shall find everywhere at Blackpool'; 'terribly mechanical-looking'; 'everything is done to avoid beauty'; 'typically joyless'; 'rather gloomy'. More space is given to the Winter Gardens than to the Town Hall, but even Joseph Emberton's Pleasure Beach Casino, which might have been expected to appeal to the international Modernist in Pevsner, is passed over in three and a half lines. Perhaps the clearest indication of the lack of 'fit' between Pevsner's expectations and the nature of Blackpool's architecture of popular pleasure occurs in his comments on the Tower: 'What a pity its base is wrapped up in a big brick building! Just remember how beautiful are the ascending curves and arches of the Eiffel Tower..., Blackpool's admitted model' (Figure 6.6). It seems unlikely that Pevsner set foot inside the Tower, and his remarks betray a startling lack of awareness of the economics and purpose of the structure: as we stated earlier, the 'big brick building' was where the money was made (Figure 6.7). It is also revealing that Pevsner missed the Grand Theatre altogether (along with – among other places – Stanley Park, the South Pier, the South Shore Open-Air Baths and the rest of Emberton's work at the Pleasure Beach), and chose to award the limited accolade of 'the architecturally best building in Blackpool' to the District Bank at the corner of Corporation and Birley Streets, which has never achieved listed building status. The key point, however, is the failure of an acknowledged enthusiast for Modernist architecture, always eager to report on pit-head baths or new secondary schools in other contexts, to find out what Blackpool had to offer in this regard in the 1960s. His lack of understanding of the nature and significance of a century of pleasure architecture in this unique setting is also remarkable in the light of his role in rekindling enthusiasm for Victorian architecture, which was gathering momentum during the 1960s.[44] This selective lack of architectural vision, which complements the failure to recognize Blackpool's importance as an industrial monument, was not confined to Pevsner and has only recently begun to be rectified.

But we must also consider what Blackpool has lost from its 'heritage of the recent past' over the last generation. The roll-call includes the Palace, originally the Alhambra of 1899, a particularly opulent 'pleasure palace' which gave way

---

44   L. Walker, '"The Greatest Century Ever": Pevsner, Victorian architecture and the lay public', in P. Draper (ed.), *Reassessing Nikolaus Pevsner* (Aldershot: Ashgate, 2004), pp. 129–48.

Figure 6.6   The Tower, which has dominated the skyline since 1894, was built five years after the Eiffel Tower, but unlike its Parisian predecessor its feet are encased in a large three-storey building by Maxwell and Tuke. The Tower was officially declared an 'Icon of England' in 2006

Source: © Blackpool Council

to a department store in 1961; the Central Station, closed in 1964, whose site still lies fallow after being earmarked for the abortive casino, and the original

**Figure 6.7   The magnificent Tower Ballroom with its Mighty Wurlitzer.
The Ballroom was built to the design of Frank Matcham and
reconstructed after a serious fire in 1956. The Tower building
also contains an Aquarium, Roof Garden and Circus; the latter
also designed by Matcham with a reservoir under the ring for
aquatic displays. The Zoo, made famous in Marriot Edgar's
comic recitation 'The Lion and Albert', no longer exists**
Source: © Blackpool Council

North Station, which was demolished ten years later; the municipal South Shore
Open Air Baths of 1923, 'an aqua arena of quite astonishing proportions and
grandiosity',[45] which was demolished in 1983; the Derby Baths of 1939, also
municipal, which were controversially lost in 1990; and on a smaller scale the
Lytham Road municipal swimming baths, demolished in 2006, and the Water
Tower of the Blackpool Sea Water Company on North Shore, which disappeared
almost unnoticed early in 2008.[46] What is significant, however, particularly on a

45   Smith, *Liquid Assets*, pp. 62–7.
46   T. Lightbown, *Blackpool: A Pictorial History* (Chichester: Phillimore, 1994); J.
Shotliff, *Images of Blackpool* (Derby: Breedon Books, 1994). For an historical background
to these events, see Walton, *Blackpool*, chapter 6.

comparative perspective, is how much was there in the first place, and how much has (in one form or another) survived, whether the British comparator is Brighton, Scarborough or Morecambe.

World Heritage Site Status

Blackpool's case for World Heritage Site status is firmly grounded in the criteria for inscription listed by UNESCO, and corresponds to the growing recognition of the importance of industrial sites and remains in Britain, beginning with Ironbridge Gorge in 1986 and accelerating in recent years with the inscription of Blaenavon Industrial Landscape (2000), Derwent Valley Mills, New Lanark and Saltaire (2001) and the Mining Landscape of Cornwall and West Devon (2006). Blackpool is a living, evolving expression of the industrial archaeology of the popular seaside holiday industry, and its pleasure palaces, promenade, boarding-houses and shows are integral to this identity. The town will seek inscription for designated sites and areas associated with the working-class holiday and entertainment industries under one or more of the cultural criteria. It could go forward under any or all of at least three of the criteria: (iii) 'a unique or at least exceptional testimony to a cultural tradition or to a civilization which is living or has disappeared'; (iv) 'an outstanding example of [an] architectural or technological ensemble or landscape which illustrates a significant stage in human history'; and/or (vi), in that it is 'tangibly associated with living traditions', in this case of regional and national popular entertainment, and the notion of 'living traditions' justifies the necessary openness to continuing change, which is essential to future survival and prosperity. Indeed the idea of continuing change that builds on living traditions is essential to the character of the place.[47]

Inscription as a World Heritage Site would formally acknowledge Blackpool's global status and prestige as the flagship site in the history of the popular seaside resort. In fact it would be the first seaside resort to enter the World Heritage List.[48] International recognition would raise the benchmark for heritage management in Blackpool. As we have seen, in the past the Council was slow to recognize the town's heritage and the years spanning the 1960s to the early 1990s saw the loss of a number of important historic buildings. Today, heritage, and World Heritage in particular, is seen as a key driver for sustainable regeneration. As a result, the present Council is actively engaged in several conservation projects encouraged specifically by the injection of grant aid from the HLF whose north-west regional office identified Blackpool as a priority area for targeted development support in 2002–04. The first of these projects saw the conservation of the Art Deco Solarium

---

47   UNESCO, *Operational Guidelines for the Implementation of the World Heritage Convention* (Paris: UNESCO World Heritage Centre, 2008), para. 77.

48   The World Heritage List currently includes 878 sites; 679 cultural, 174 natural and 25 mixed; http://whc.unesco.org/en/list (accessed 1 August 2008).

at Harrowside on the south promenade, opened in 1938, and transformed in 2004 as an environmental centre of excellence. The HLF has also supported projects at the Grand Theatre (now re-branded as the National Theatre of Variety), St John's Church (the original Blackpool parish church) and elsewhere, and is currently funding the enhancement of the historic core of the town through its Townscape Heritage Initiative (THI) and the Stanley Park regeneration project.[49] Aspiring to and obtaining World Heritage Site status are often triggers for increased inward investment by helping to lever additional sponsorship and enhancing the chances of grant aid for new projects. As well as further conservation work, including an extension to the THI area, Blackpool's ambitions are seemingly boundless. A number of heritage-related attractions are planned, including the relocation from London of the National Theatre Museum set to become the National Museum of Performing Arts.

Heritage is a prized commodity in increasingly competitive tourism markets, with the choice of destination increasingly influenced by heritage provision. Blackpool already attracts some 10 million visitors every year, considerably more than existing World Heritage Sites.[50] The town's enhanced status as a heritage destination would provide significant additional tourism opportunities and may help attract a new, different, more up-market visitor base nationally and internationally. Heritage is also increasingly seen as an indicator of local prestige and well-being. Building a successful case for World Heritage Site inscription can often raise civic pride in an area and play a part in social regeneration. In the case of Blackpool, parts of which have suffered from extreme deprivation in recent years, a place on the World Heritage map would promote a positive image of the town and a clear focus for community cohesion. Putting together the bid will require the engagement of as many people as possible in helping to determine the value and significance of the place. Participation in heritage projects and in local events, perhaps linked to national events like Heritage Open Days, will be important to get the community engaged in working together to discover, present and conserve the town's heritage. Particular emphasis will be given to involving young people as both participants and heritage 'ambassadors', with perhaps a World Heritage Discovery Day being staged in conjunction with local schools.[51] A crucial aim will be to encourage people to tell their own stories, to share their personal and often intangible heritage and 'unofficial' history of Blackpool, to explore further the forces that link these memories to specific places, and thus hold together what the Council is seeking to identify, mark and celebrate under the World Heritage banner.

49   http://www.blackpool.gov.uk/Services/S-Z/TownscapeHeritage/Home.htm accessed 28 April 2008; http://www.friendsofstanleypark.org.uk (accessed 28 April 2008).
50   See above, note 17.
51   UNESCO, *World Heritage in Young Hands* (Paris: UNESCO, 1999); UNESCO, *Mobilizing Young People for World Heritage*, World Heritage Papers 8 (Paris: UNESCO World Heritage Centre, 2003).

It is important to note that World Heritage Site status need not be at odds with the much-needed regeneration work. This is because Blackpool would be put forward under the category of 'cultural landscape'. Cultural landscapes are seen as retaining 'an active social role in contemporary society closely associated with the traditional way of life, and in which the evolutionary process is still in progress' while at the same time exhibiting significant material evidence of their evolution over time.[52] UNESCO recognizes that World Heritage Sites falling under this category are living and working places that must be allowed to grow dynamically and organically in response to their local environment and society as they have in the past. The Masterplan vision for the 'New Blackpool' can be seen as an example of the town's continuing evolution in response to economic and social change. It is here, together with the 'Admission All Classes' project, that the cultural landscape category comes into its own. The crucial concerns, however, will be to ensure that the next phase of Blackpool's regeneration adds to the heritage and cultural traditions of the place, rather than erasing them, and to overcome the hostile prejudices of important sections of the British media.[53]

Conclusion

We have demonstrated that Blackpool has no credible challenger for the title of world's first working-class seaside resort. The town pioneered popular tourism in the nineteenth century and has continued to adapt to the changing desires of the holiday market while retaining and celebrating much of its unique cultural landscape, and sustaining its distinctive atmosphere of revelry, participation, fun and excitement. Today Blackpool constitutes a meeting point and melting pot of contested and contradictory spaces; a living, evolving expression of the archaeology of the popular seaside holiday and entertainment industry; always in flux but always retaining a core identity and ambience, with an impressive array of surviving architectures and built environments dedicated to the provision of leisure and enjoyment.

For those harbouring traditional notions about the content and nature of 'heritage' Blackpool poses a difficult but fascinating challenge. Until recently the town was rarely associated with the word 'heritage', but aversion to this apparently incongruous juxtaposition is slowly beginning to change with widening recognition within the heritage sector of the economic and cultural significance of the seaside in British society and beyond. However, when the local Council first revealed its ambitious, but not unreasonable, plans to put Blackpool forward as a potential World Heritage Site, many in the heritage sector, in tune with initial media reaction, considered the proposal audacious and counter-intuitive.

---

52   UNESCO, *Operational Guidelines*, annex 3, 10, ii.
53   See above, note 3.

In making a case for World Heritage Site status, such a unique site will require a unique kind of bid; one that embraces changing perceptions of and conflicts around the 'heritage of the recent past', one that further stimulates the overlap between heritage studies and popular culture, and one that squarely confronts the mind set that sees Blackpool as the antithesis of heritage.

References

Band, B. *Blackpool Grand Theatre 1930–1994*. Lytham: Barry Band, 1994.

Beatty, C. and Fothergill, S. *The Seaside Economy*. Centre for Regional Economic and Social Research, Sheffield Hallam University, Sheffield, 2003a.

Beatty, C. and Fothergill, S. *A Case Study of Blackpool*, Seaside Town Research Project, Paper No. 3. Centre for Regional Economic and Social Research, Sheffield Hallam University, 2003b.

Blackpool Council. *New Horizons: Blackpool Resort Masterplan*. Blackpool, 2003.

Blackpool Council. *Vision for Blackpool: Blackpool's Community Plan 2004–2020*. Blackpool, 2004.

Blackpool Council. *Towards a World Class Resort Destination: Submission to the Casino Advisory Panel*. Blackpool, 2006a.

Blackpool Council. *Heritage Strategy 2006–10*. Blackpool, 2006b.

Brodie, A. and Winter, G. *England's Seaside Resorts*. Swindon: English Heritage, 2007.

Casino Advisory Panel. *Final Report of the Casino Advisory Panel*. London, 2007.

Cross, G. and Walton, J. K. *The Playful Crowd: Pleasure Places in the Twentieth Century*. New York: Columbia University Press, 2005.

Fairley, A. *De La Warr Pavilion: The Modernist Masterpiece*. London: Merrell, 2006.

Gray, F. *Designing the Seaside: Architecture, Society and Nature*. London: Reaktion, 2006.

Guise, B. and Brook, P. *The Midland Hotel: Morecambe's White Hope*. Lancaster: Palatine Books, 2007.

Lightbown, T. *Blackpool: A Pictorial History*. Chichester: Phillimore, 1994.

Meller, H. *European Cities 1890–1930s*. Chichester: John Wiley and Sons, 2001.

Morris, S. 'Forget the Taj Mahal and Pyramids: Blaenavon is the Place to Visit', *The Guardian*, 26 March, 2008.

North West Regional Development Agency. *Blackpool: An Action Plan for Sustainable Growth. Report of the Blackpool Task Force*. Blackpool, 2007.

Pevsner, N. *The Buildings of England: North Lancashire*. Harmondsworth: Penguin, 1969.

ReBlackpool. *Blackpool Resort Masterplan: Rethink, Rebuild, Renew*. Blackpool, 2007.

Russell, D. 'Selling Bradford: Tourism and northern image in the late twentieth century', *Contemporary British History* 17, (2003): 49–68.

Shotliff, J. *Images of Blackpool*. Derby: Breedon Books, 1994.

Simon, B. *Boardwalk of Dreams*. New York: Oxford University Press, 2004.

Smith, J. *Liquid Assets: the Lidos and Swimming Pools of Britain*. London: English Heritage, 2005.

Smith, M.K. (ed.) *Tourism, Culture and Regeneration*. Wallingford: CABI, 2007.

Stofik, M.B. *Saving South Beach*. Gainesville: University Press of Florida, 2005.

UNESCO. *World Heritage in Young Hands*. Paris: UNESCO, 1999.

UNESCO. *Mobilizing Young People for World Heritage*, World Heritage Papers 8. Paris: UNESCO World Heritage Centre, 2003.

UNESCO. *Operational Guidelines for the Implementation of the World Heritage Convention*. Paris: UNESCO World Heritage Centre, 2008.

Walker, B. (ed.) *Frank Matcham, Theatre Architect*. Belfast: Blackstaff Press, 1980.

Walker, L., '"The Greatest Century Ever": Pevsner, Victorian architecture and the lay public', in *Reassessing Nikolaus Pevsner*, edited by P. Draper, 129–48. Aldershot: Ashgate, 2004.

Walton, J.K. 'The demand for working-class seaside holidays in Victorian England', *Economic History Review* 34, (1981): 249–65.

Walton, J.K. 'The world's first working-class seaside resort? Blackpool revisited, 1840–1974', *Transactions of the Lancashire and Cheshire Antiquarian Society* 88, (1992): 1–30.

Walton, J.K. 'The Blackpool landlady revisited', *Manchester Region History Review* 8, (1994): 23–31.

Walton, J.K. *Blackpool*. Edinburgh: Edinburgh University Press, 1998.

Walton, J.K. *The British Seaside: Holidays and Resorts in the Twentieth Century*. Manchester: Manchester University Press, 2000.

Walton, J.K. *Tourism, Fishing and Redevelopment: Post-war Whitby, 1945–1970*. University of Cambridge: Institute of Continuing Education, Occasional Paper No. 5, 2005.

Walton, J.K. *Riding on Rainbows: Blackpool Pleasure Beach and its Place in British Popular Culture*. St Albans: Skelter Publishing, 2007.

Walton, J.K. and Wood, J. 'World heritage seaside', *British Archaeology* 90, (2006): 10–15.

Walton, J.K., and Wood, J. 'History, heritage and regeneration of the recent past: The British context', in *Interpreting the Past V, Part 1: The Future of Heritage: Changing Visions, Attitudes and Contexts in the 21ˢᵗ Century*, edited by N. Silberman and C. Liuzza, 99–110. Brussels: Province of East-Flanders, Flemish Heritage Institute and Ename Center for Public Archaeology and Heritage Presentation, 2007.

Walton, J.K. and Wood, J. '"La Station Balnéaire Comme Site du Patrimoine Mondial? Le Cas de Blackpool', in *Les Villes Balnéaires d'Europe Occidentale du XVIIIe Siècle à Nos Jours*, edited by Y. Perret-Gentil, A. Lottin and J.-P. Poussou. Paris: Presses de l'Université Paris-Sorbonne, 2008.

Chapter 7

Values not Shared: The Street Art of Melbourne's City Laneways

Tracey Avery

In late June 2008, a feature story in *The Age* newspaper, Victoria's broadsheet daily, led with the headline, 'Anti-graffiti lobby sees red at heritage-listing proposal', and the opening sentence 'Considered by many as vandalism, some of the graffiti of Melbourne's laneways could receive heritage protection in a move supported by the National Trust and Heritage Victoria'.[1] Despite the fact that neither organization had proposed to heritage list graffiti, one prompt for the story was the announcement that readers of the *Lonely Planet* guidebooks had voted the street art of Melbourne's laneways one of '...the nation's most popular cultural attractions'.[2] The British graffiti artist Banksy commented that 'Melbourne's laneways were "arguably Australia's most significant contribution to the arts since they stole all the Aborigines' pencils" (sic)', thereby giving international recognition to Melbourne's street art.[3] In the same article, the National Trust of Australia (Victoria) (NTAV), commented that some of the work in particular laneways may be worth documenting, such as Hosier Lane, where the local authority, Melbourne City Council had commissioned art work.[4]

The publication of this article, which included positive and negative comments from a range of Government, business and community groups, led to a flurry of local and national media interest. The National Trust received a broad range of critical commentary, but most particularly the accusation that a respected heritage organization was 'condoning vandalism'. The spectrum of responses to the values ascribed to this very public presence in Melbourne's urban environment – from congratulations on discussing the topic, to the resignation of a member (one of 18,000) – illustrates well the complexity of cultural landscapes; all contain the acceptable and the subversive.[5]

---

1   Cameron Houston, 'Anti-graffiti lobby sees red at heritage-listing proposal', *The Age*, June 23, 2008: 5.
2   Ibid.
3   Ibid.
4   Ibid.
5   The categories of 'graffiti art' and 'street art' might both be understood as having artistic merit for some but the term 'street art' is preferred by artists who paint in spaces where they have permission to do so; the term 'graffiti' is used for illegal work.

The Burra Charter is the framework which informs the NTAV's assessments of heritage value. According to this framework aesthetic, historic, scientific and social value should all be considered in the designation of a place or object's cultural significance.[6] In an extended definition of cultural values, Article 13 of *The Burra Charter* (Co-existence of cultural values) states that the 'co-existence of cultural values should be recognized, respected and encouraged, especially in places where they conflict.'[7] The explanatory notes to Article 13 read as follows:

> For some places, conflicting cultural values may affect policy, development and management decisions. In this article, the term cultural values refers to those beliefs which are important to a cultural group, including but not limited to political, religious, spiritual and moral beliefs. This is broader than values associated with cultural significance.[8]

The value of those practices is different to different sectors of the wider community. While the values may be subjective or difficult to define, it is only when the existence of places or practices are threatened or celebrated that their current heritage values tend to be articulated by both the public and heritage professionals. The purpose of this discussion of heritage value in aspects of Melbourne's urban environment is to tease out the conflicts and contradictions inherent in the recognition and framing of discrete elements as 'heritage' in cultural landscapes.

The National Trust's Role in the Protection of Melbourne's Physical and Intangible Cultural Landscape

Following the National Trust model in the United Kingdom, a local incorporation of the Trust was established in the state of Victoria in 1956. The primary purpose of the Victorian Trust was to advocate for the protection of the state's historic places. From 1959 the NTAV took on the management of historic sites and opened them to the public. Since that time, the organization has always maintained a strong advocacy arm. The NTAV has documented and classified buildings, industrial infrastructure, public art, trees, gardens and landscapes, as well as more specialized objects and sites such as pipe organs and cemeteries. There are currently over 7,000 classifications, each graded according to local, regional, state, national or international significance. Although the Trust's classification system does not legally protect a place from alteration or demolition, the expertise embodied in the

---

6   Australia ICOMOS, *The Burra Charter: The Australia ICOMOS Charter for Places of Cultural Significance*, (Victoria: Australia ICOMOS, 1999).

7   Peter Marquis-Kyle and Meredith Walker, 'Articles and explanatory notes', in *The Illustrated Burra Charter: Making Good Decisions about the Care of Important Places*, (2nd ed.), Australia ICOMOS, Burwood, Victoria, 2004, 106.

8   Ibid.

creation of the register of classifications was enshrined in the eventual creation of statutory protection for heritage through the state planning system, most recently in the *Victorian Heritage Act* (1995).[9]

In recent years the increasing role of Government bodies in both defining and monitoring the listing and development of heritage places, has forced the NTAV to re-examine the best uses of its classification system as a tool for its advocacy voice. Alongside a reduction in the number of new classifications, there has been an increased focus on the promotion of items and places of 'future' heritage significance; many of these have intangible characteristics based on social value.[10] In 2006 and 2007, the NTAV, in association with the *Herald Sun* and *The Age* newspapers, ran a 'Heritage Icons' competition for the public to nominate their favourite Victorian icons.[11] Winners included Phar Lap (Australia's most famous racehorse), Australian Rules Football, and the Skipping Girl neon sign in Melbourne (see Figure 7.1). The awarding of icon status did not automatically generate a heritage listing, and indeed this aspect of the icons program was important as it allowed the Trust to recognize and celebrate heritage without the fear of potential listing which exists in some sectors of the community.[12] However, the Skipping Girl neon sign was listed by the NTAV and was subsequently entered on the Victorian Heritage Register. The merits of the icons program have been twofold: firstly they allow the NTAV to canvas public opinion for places and objects which the community value and secondly, listing can be treated as a separate issue.

On the other hand, consideration of the listing of Australian Rules Football exposed the practical difficulties of listing an intangible cultural practice which contains tangible places and artefacts. Heritage listing operates within Government planning regulations and relies heavily on the ability to define the physical boundary of a place. Listing can be recommended as a practical and positive step, as in the example of the Skipping Girl neon sign. Alternatively, listing can be rejected as a conservation tool when the planning implications of listing may prove counterproductive to the survival of that cultural artefact or endeavour. In terms of the positive recognition of heritage, it seems that the NTAV's unique independent ability to denote heritage significance has meant that some public

---

9   State Government of Victoria, *Heritage Act 1995 (no. 93)*, (Victoria: State Government of Victoria, 2007). This statutory protection was first introduced as the *Historic Buildings Act* (1974).

10   In this chapter I use Chris Johnston's definition of social value. See, Chris Johnston, 'Social values – overview and issues', in *People's Places: Identifying and Assessing Social Value for Communities*, (Canberra: Australian Heritage Commission, 1994a) and Chris Johnston, *What is Social Value?* (Canberra: Australian Government Printing Service, 1994b).

11   Other Australian National Trusts have run similar icon programs, including the South Australia and Queensland National Trusts.

12   See Hannah Lewi and Stephen Neille for further discussion of anti-listing community sentiment, 'Drawing in and fading out: The case of the Commonwealth Games housing village, Western Australia', *Journal of Architectural Education* (2007): 15–21.

Figure 7.1   Skipping Girl neon sign, 1970 (replacement of a 1930s sign, Abbotsford, Melbourne, Victoria, photograph circa 2006)
Source: © National Trust of Australia (Victoria)

groups and individuals have increasingly looked to the Trust to list 'new' categories of heritage. In 2006, the Melbourne Football Club, Australia's oldest club in the Australian Football League (AFL), approached the NTAV to make a nomination for classification. While the Trust welcomed the interest generated by the proposal, practical concerns arose over what might be included in a listing, such as the design of football jerseys and the likelihood that other clubs might also seek listing. Practical problems could arise when one club wanted to change its jersey designs and colours, move away from its traditional stadium, merge with another club or change rules in the game. Could the NTAV listing be used to prevent changes to those items and activities noted at the time of the listing, as would be the case with an object or place? The reasons given in favour of listing revolved around the desire to recognize the Melbourne Football Club as the first club, dating back to 1858.

To broaden the debate and seek public and other professional views, the Trust's Cultural Heritage Advisory Committee supported the staff's recommendation that a symposium be held on the subject of intangible cultural heritage. The resulting symposium, *Intangible Cultural Heritage: A New Field of Endeavour*, was held in association with The Cultural Heritage Centre for Asia and the Pacific at Deakin

University and Australia ICOMOS, in Melbourne in July 2008.[13] The symposium featured local and international speakers who discussed a wide range of intangible cultural heritage, including cultural practices of indigenous Australians, Jewish heritage, children's play, and sport. The latter was represented in a paper by Robert Pascoe who spoke on the history and significance of the AFL.[14] In the course of the final plenary session, the Trust concluded that it was not able to list the Melbourne Football Club, but given the Club's expressed interest in acting as a case-study for recognition as intangible cultural heritage, the Trust would seek to refer the case to the Commonwealth Government in discussions over Australia's possible adoption of the 2003 UNESCO *Convention for the Safeguarding of the Intangible Cultural Heritage.*[15]

Blurring the Boundaries of 'Acceptable' Cultural Value: Graffiti Versus Street Art

The article in *The Age* on Melbourne's graffiti and street art mentioned in the introduction to this chapter coincided with the promotion of the intangible cultural heritage symposium and the symposium was mentioned in the same article.[16] The Trust was well known as an early supporter of street art, having adopted a national listing, in 1997, for a Keith Haring mural in Collingwood – the artist's only surviving work in Melbourne (see Figure 7.2). This was a clear example of a commissioned art work and not problematic in terms of its legal place in the urban landscape. While the extensive tagging of public and private property is unquestionably illegal the representation of a small proportion of street art and political graffiti as popular culture through art publications and exhibitions has blurred the boundaries between art and vandalism. In this section I will consider how different perceptions of the validity of graffiti and street art has made even legal street art work difficult to formally recognize as heritage.

---

13   The National Trust of Australia (Victoria), The Cultural Heritage Centre for Asia and the Pacific at Deakin University, and Australia ICOMOS, *Intangible Cultural Heritage: A New Field of Endeavour*, Melbourne, 1–3 July 2008, unpublished conference proceedings.

14   Robert Pascoe, 'Sport: The unbearable lightness of a bag of wind' in *Intangible Cultural Heritage: A New Field of Endeavour*, ibid.

15   UNESCO, *Convention for the Safeguarding of the Intangible Cultural Heritage*, (Paris: UNESCO, October 17, 2003), http://unesdoc.unesco.org/images/0013/001325/132540e.pdf (accessed November 3, 2008). The Australian Council of National Trusts recently submitted a response to the Commonwealth Government's call for submissions on the possible impact of Australia becoming a signatory to this convention, Australian Council of National Trusts, Letter to Assistant Secretary Literature and Indigenous Culture, Commonwealth Department of the Environment, Water, Heritage and the Arts, September, 2008.

16   Houston, 'Anti-graffiti lobby sees red at heritage-listing proposal', *The Age*, June 23 (2008): 5.

Figure 7.2   Untitled (informally known as 'The Collingwood Mural'), Keith Haring, Collingwood, Melbourne, Victoria
Source: © National Trust of Australia (Victoria) (photograph circa 1997)

Despite the recent shock waves caused by references to the graffiti of Melbourne's laneways as heritage, debates about heritage listing and graffiti in Melbourne's cultural landscape were explored in the 1990s when the NTAV drafted a classification for a piece of political graffiti. This was known as the 'Keon graffiti' (see Figure 7.3). This graffiti, which consisted of the slogan 'Keon traitor to ALP', referred to Stan Keon, the State and subsequently Federal member covering the inner Melbourne suburb of Richmond who in 1955 defected from the Australian Labor Party (ALP) to the newly-formed Democratic Labour Party (DLP).[17] A meeting of the Trust's Twentieth Century Buildings Committee in May 1998 resolved that 'a *prima facie* case existed for the classification at the local level'.[18] An extract from the draft statement of significance reads as follows:

> The 'Keon traitor to the ALP' graffiti, painted between 1955 and 1961, is
> historically and socially significant at the local level. The graffiti is the only

---

17   National Trust of Australia (Victoria), Draft Classification Report for 'Keon Graffiti', Tanner Street, Richmond, File number 6969, November, 1998.
18   National Trust of Australia (Victoria), Minutes of Twentieth Century Buildings Committee, item 6 (h), Temporary file 'Keon Graffiti', Tanner Street, Richmond, April 8, 1998, 2.

Figure 7.3   Keon Graffiti, Richmond, Melbourne, Victoria
Source: © National Trust of Australia (Victoria) (photograph circa 1999)

tangible reminder in the Richmond area of the great 'split' of the Democratic Labor Party from the Australian Labor Party which occurred in 1955. This split turned the suburb of Richmond, a heartland of labour politics, into 'one of the bitterest and most damaged battlegrounds'. The graffiti demonstrates the depth of emotion in the streets during this period.[19]

The reference to 'depth of emotion' being a driver for the creation of this political comment on an urban factory wall indicates that the Trust was aware of graffiti as an important marker of history in the urban cultural landscape. However, by identifying this graffiti as 'heritage' the Trust exposed a range of strong community opinions about whose heritage should be recognized and preserved.

Initial media attention directed at the Trust's recognition of the 'Keon graffiti' indicated the polarization of opinion over graffiti in general. For instance, *The Age* editorialized 'Some call it vandalism; others call it art. To the National Trust, graffiti is history'.[20] From other quarters there was mild admiration for the Trust's foresight in acknowledging '...the sometimes informal, even seditious, nature of

---

19   National Trust of Australia (Victoria), Draft Classification Report for 'Keon Graffiti', Tanner Street, Richmond, File number 6969, 1.

20   Penny Fannin, 'Slogan's Heroes – trust fights for history writ large', *The Age*, April 12 (1999): 3.

the historical artefact...'.[21] However, within a fortnight of the first media publicity, it was reported that the sign had been defaced by a large swastika.[22] At the time the Trust's conservation officer, Rohan Storey, was quoted saying that '...he was not surprised by the new addition. "It was inevitable. You have to accept something like that happening"'.[23] In the event the classification was not formally adopted. The case of the 'Keon graffiti' illustrates some of the problems at the heart of assigning heritage value to an ephemeral result of a counter-cultural practice, namely the preservation of a contested practice in a contested space.

A few years prior to the NTAV's listing of the Keith Haring mural, a series of newspaper articles drew attention to the vexed issues around its historic significance and increasing deterioration; these articles formed an important research component of the subsequent classification report.[24] For instance, art academic Chris McAuliffe, expressed his view regarding the impermanence of this work, arguing that '...as graffiti, it should be left to fade... If you subject it to conservation procedures then you transpose graffiti into a realm that it was opposed to. You make it art'.[25] Yet, the production of the mural itself was by invitation from the higher education college who owned the site (the Northern Metropolitan College of TAFE) and Haring was assisted by students in the creation of the work *as art*.[26] The vexed conservation issue of future re-touching due to fading was considered in the statement of significance:

> Crucial to the fate of the mural and, given its exposure to the elements, is whether the artist himself would have accepted the deterioration of the mural or have condoned some form of restoration. Haring's own feelings appear to have been ambivalent in the matter. In favour of restoring the mural i.e., repainting – is the fact that the simplistic three colour design devoid of subtle harmonies would not present serious problems in restoring it to its original condition. Opinion appears to be divided regarding the moral considerations in the matter and even the Estate of Keith Haring is unclear in this matter. Work has already been done to prevent further erosion of the mural's wall.[27]

The local Council (City of Yarra) successfully nominated the mural to the Victorian Heritage Register in 2004. The work has been left to fade but recent reports indicate

---

21 Andrew Masterson, '...And so it was written', *The Age*, April 16 (1999): 16.

22 'Paint-can Curse Gets Historic Daubing', *Melbourne Times*, April 21 (1999): 3.

23 Ibid.

24 Andrew Masterson, 'Off the wall art', *The Age*, December 27, *Summer Age* supplement, (1994): 4–5.

25 Ibid.

26 Ibid.

27 National Trust of Australia (Victoria), Classification Report for 'Keith Haring Mural', Johnston Street, Collingwood, File number 6675. Extract from Statement of Significance, 4 August 1997.

that Haring later said that the images could be touched up in the future.[28] Although most views reported indicate that street art is ephemeral and therefore, should not be preserved indefinitely, the listing of Haring's work indicates some value of ephemera to the wider cultural landscape. The remainder of this chapter will explore the nature of the competing values which need to be negotiated if street art is to be recognized and managed as heritage in urban cultural landscapes.

Competing (for) Values in the Urban Cultural Landscape

The laneways of Melbourne's CBD are a series of narrow lanes and alleyways which run between the city's grid-like street layout. Street and installation art in the laneways has signified their gentrification from their former status as 'backwaters, populated by drunks and rubbish bins'.[29] Changes in the licensing laws in the 1990s saw the proliferation of 'hole in the wall' bars, cafés and restaurants, bringing these parts of the city to life at night. Together with a dramatic increase in the residential population of the CBD, street art and graffiti have grown alongside these trends. An extract from the City of Melbourne's 2008 strategy plan *Future Melbourne* summarizes the City's view of the interconnection between its physical and cultural features:

Melbourne has also expanded rapidly as an entertainment and cultural centre with an increase in bars, restaurants, clubs, galleries and cultural facilities. There are 1,500 bars, cafés and restaurants in the central business district area alone and a multitude of entertainment, cultural and dining venues in its inner suburbs.

The CBD's greatest strengths lie in its number of heritage buildings, pedestrian scale and tree lined streets built on the Hoddle Grid which laid out our pattern of streets and laneways in 1837. *Its streets, laneways and other public spaces are full of life and intrinsic to Melbourne's physical character* [my italics].

The CBD is also a popular residential address and residents make an important contribution to the city. In 2008, there were 18,000 residents in the CBD.[30]

---

28   Greg Burchall, 'Artist's legacy fades but not forgotten', *The Age*, June 12, 2008: http://www.theage.com.au/news/film/bfilmb-artist-keith-harings-legacy-fades-but-is-not-forgotten/2008/06/11/1212863732691.html (accessed November 3, 2008). 'I asked Keith a few times, in later years, what he wanted to do with it and he said, "Get some sign writers in, freshen it up". There was absolutely no preciousness to it; that would have been counter to the spirit of everything', quoting gallery owner and art consultant John Buckley who commissioned the Collingwood mural.

29   Jewel Topsfield, 'Brumby slams Tourism Victoria over graffiti promotion', *The Age*, October 1 (2008): http://www.theage.com.au/articles/2008/10/01/1222651140951. html (accessed October 3, 2008). Quoting Paul Round, graffiti project worker.

30   City of Melbourne, *Future Melbourne*, 2008: http://www.futuremelbourne.com. au/wiki/view/FMPlan/S1dGrowthAndStrategicAreas (accessed on October 27, 2008).

At the state and local government levels, which have responsibility for tourism, planning and arts, graffiti and street art have pushed internal tensions into the public arena. At the time of writing the state Government department, Tourism Victoria, was criticized by the state Premier, John Brumby, for promoting 'graffitied lanes in a recreated cityscape of Melbourne at Florida's Disney World' where a food and wine expo was being held in the United States.[31] The Premier's comments illustrated the contested nature of the laneways, for him the good aspects of the laneways were its '"European-type" style…openness …little restaurants. I don't think graffiti is what we want to be displaying overseas. We've put through very tough laws to discourage graffiti – it's a blight on the city'.[32] The Premier's comment was countered by a range of comments, including from arts workers and the Mayor of an adjoining inner-city council, raising the suggestion that popular public opinion supported the commissioning of street art in the laneways, as Andrew MacDonald, Melbourne's City Lights public art project director commented,

> Whether … [Government ministers] like it or not, Melbourne's street art is a big hit with tourists. They are behind the times and out of step with public opinion – I say bravo to Tourism Victoria for trying to express something which is genuinely happening here.[33]

Street art, graffiti, cafés and bars, have together contributed to a cultural landscape in the laneways which is very much contested and publicly debated. Other groups argue that the public is against graffiti and street art, saying that there is no 'difference between the graffiti in Hosier Lane to 'tagging', where artists leave behind their name or representative symbol'.[34] Others make the point that the cost to councils across Australia for graffiti *removal* has been estimated at (AU) \$260 million.[35]

Colloquial definitions of graffiti and street art are predicated on the legality of the work. In Melbourne there is competition among artists to gain access to permitted spaces. Throughout Melbourne's laneways there is also evidence of competition for space amongst illegal works. Consequently, it can be difficult for the wider community to distinguish between legal street art and illegal graffiti. Spaces like Hosier Lane where permission can be sought to apply new art works, are not immune to being filled with illegal work. Anti-graffiti campaigns include 'ban the can' and city walls, including Hosier Lane, have been plastered with

---

31   Jewel Topsfield, 'Brumby slams Tourism Victoria over graffiti promotion', op. cit.
32   Ibid.
33   Ibid.
34   Rachel Brown, 'Melbourne Graffiti considered for Heritage Protection, *The World Today*, ABC Radio, June 23, 2008: http://202.6.74.88/news/stories/2008/06/23/2282978. htm?site=melbourne (accessed October 3, 2008), quoting Scott Hilditch, chief executive of Graffiti Hurts Australia.
35   Ibid.

posters warning of $500 fines for graffiti makers.[36] These civic efforts contradict the Government's promotion of the mixture of graffiti and street art in the laneways in, for instance, tourism campaigns such as the previously mentioned Tourism Victoria re-creation of Melbourne's laneways including graffiti. But can the inherent contradictions, including legality, be managed to take account of the wide range of competing community values?

As the discussed examples make clear the graffiti and street art in the laneways is the subject of vociferous debate. However, it may transpire that heritage values have a role to play in enabling the community to openly discuss the issues concerning street art and graffiti, rather than simply perceiving heritage protection mechanisms as likely to stultify the legal practices.

Preserving Ephemeral Cultural Landscapes?

In these most recent debates on the city's graffiti, supporters of street art, such as the City Council, again made the distinction between street art and 'graffiti, vandalism and tagging'.[37] However, these categories – legal and illegal – are far from clear. For instance, recently, a small work in Cocker Alley by the British graffiti artist, Banksy, was covered by a sheet of Perspex to 'prevent it being vandalized' (see Figure 7.4).[38] Ironically, this work was put up illegally and was then preserved by Melbourne City Council.[39] Not surprisingly, the Perspex suffers from tagging. Some large areas of other walls of street art in the city have been treated with a clear lacquer to preserve work. While this does not prevent them from being covered in new work, these walls are repeatedly cleaned. Artists themselves have questioned both the cost of cleaning and the futility of preventing new layers.[40]

While heritage bodies may consider the longer-term viability of listing sites of street art, neither supporters nor detractors of the practice seek any such recognition. When the subject of heritage listing in Melbourne's laneways arose in June 2008, the chief executive of the organization Graffiti Hurts Australia was reported as saying that 'protecting Melbourne's laneway decorations would send a dangerous

---

36   'Ban the can' refers to calls to ban the sale of paint in spray cans.
37   City of Melbourne, *Future Melbourne*, 2008: http://www.futuremelbourne. com.au/wiki/view/FMPlan/S2G5Creative (accessed October 27, 2008). Also see City of Melbourne, *Graffiti Management Plan*, Melbourne, 2006: http://www.melbourne.vic.gov. au/rsrc/PDFs/Graffiti/graffitti_management_plan_2006.pdf (accessed November 3, 2008).
38   Jewel Topsfield, 'Brumby slams Tourism Victoria over graffiti promotion', op. cit.
39   Ibid. See also City of Melbourne's list of approved street art, where the Banksy stencil was subject to a retrospective planning application (STA 020) and approved, http:// www.melbourne.vic.gov.au/info.cfm?top=145&pa=3274&pg=3842 (accessed November 3, 2008). The work was subsequently vandalized, http://www.theage.com.au/national/ the-painter-painted-melbourne-loses-its-treasured-banksy-20081213-6xzy.html (accessed March 3, 2009).
40   Anonymous collective of street artists, personal communication October 1, 2008.

Figure 7.4   'Little Diver', Banksy, 2003, Cocker Alley, Melbourne
Source: © National Trust of Australia (Victoria) (photograph October 2008)

message that graffiti is acceptable and open the floodgates to vandalism'.[41]
However, a supporter of street art also argued that it was inappropriate to heritage
list graffiti, saying,

> The work is ephemeral. It's not meant to last. It lasts purely as long as the
> weather and other graffiti artists allow it to last... When you interfere with what

---

41   Rachel Brown, 'Melbourne Graffiti considered for Heritage Protection', op. cit.

is an organic process like that, you actually make the graffiti stagnant and what makes graffiti thrilling and interesting to the public and to other graffiti artists is the fact that it's a never-ending, changing, kind of living art form.[42]

As Cultural Heritage Manager at the NTAV I have commented on the difficulties with listing ephemeral work,

> We recognize that it may not be possible to list graffiti for the long-term because of it's ephemeral nature,...So it may be that we end up saying that what the best thing to do is take proper records of it and interview artists and take public comments...that ...becomes a visual and oral history about graffiti. But we may not be able to protect the individual pieces themselves.[43]

However, given the links made between graffiti, street art and vandalism the Trust received criticism for making even these carefully circumscribed comments on conservation and the value of street art.[44]

Professional photography has long been used to document street art. In New York in the 1980s, the movement of art from artists' studios to their surrounding streets in areas like SoHo caught the attention of photographers and the art world generally.[45] The brief life of most works was accepted and publications and photographic exhibitions on graffiti and street art have proliferated since seminal texts were published in the 1980s such as *Subway Art* and *Spraycan Art*.[46] Recently the Trust has been approached by a number of professional photographers with an interest in documenting street art in Melbourne's laneways. Following the principles of *The Burra Charter*, documentation is an essential step in the conservation process, which aids the assessment of significance and prevents the total loss of the work from the historical record.[47] Additionally, it is the least contested heritage management activity. While the Trust has received offers to assist with the documentation of street art in the laneways, no decision has yet been made as to whether the Trust would consider undertaking this work.

42   Ibid., quoting Andrew Mac.
43   Ibid., quoting Tracey Avery.
44   It should be noted that the Trust followed up these comments with a press release re-stating that it did not condone tagging and that these particular comments concerned legally made street art in Hosier Lane. Whereas the terms graffiti and graffiti art seem to have been used in the past to include street art, the strength of community concern over tagging led us to re-state our position referring only to street art.
45   David Robinson, *Soho Walls: Beyond Graffiti*, (London: Thames and Hudson, 1990), 12–13.
46   Martha Cooper and Henry Chalfant, *Subway Art*, (London/New York: Thames and Hudson, 1984) and Henry Chalfant and James Prigoff, *Spraycan Art*, (London/New York: Thames and Hudson, 1987).
47   Australia ICOMOS, *The Burra Charter: The Australia ICOMOS Charter for Places of Cultural Significance*, (Victoria: Australia ICOMOS, 1999).

Conclusion

The initial problem put to the National Trust could be summarized as follows: Is street art now part of our shared cultural heritage and should it be validated by heritage listing? As Hannah Lewi and Stephen Neille have noted on the role of documentation in the heritage process, even the recording of a contested object or practice can precipitate its destruction because not all sectors of the community may agree that an item is worth preserving for the future.[48] This was the Trust's own experience when it reported its consideration of the heritage significance of the 'Keon graffiti'. The push amongst heritage professionals to record and preserve more recent examples of buildings, objects and cultural practices creates the dilemma of trying to preserve something as cultural heritage before it is lost but not before it has achieved some community recognition as heritage. This is especially true of so-called 'everyday' practices where documenting equates to 'raising its significance, (and thus) its status is inevitably intervened in and altered'.[49] Moreover, in Lewi and Neille's appraisal of their own role as researchers their 'attempt ... to remain distant from heritage advocacy was at best hopeful and at worst naïve.'[50] In the NTAV we have a strong record of being at the forefront of recognizing recent heritage; some community backlash is inevitable when new ideas are being explored. Ultimately, as heritage professionals we need to consider how this recent heritage can remain part of the cultural landscape without threatening other community values to the extent that the effect of our actions is to hasten the destruction of the very item or site we wish to save. Consequently, I argue that we have to engage with community debate rather than remain outside of it so that communities no longer see heritage as a blunt preservation order. As the icon programs have shown, cultural heritage can be a dialogue and a celebration of living cultural practices.

The Trust has not yet developed a policy on street art. In the current climate we are too close to the present unfolding of public opinion and Government policies to intervene. At this stage we can listen, record and take note of the range of community views. The intensity of public debate surrounding this topic will undoubtedly read as having historic social significance at some future date. Whether current heritage protection processes would still be adopted in the future to recognize surviving work remains to be seen.

Acknowledgements

I would like to thank Celestina Sagazio, Senior Historian at the NTAV for her assistance and comments in the preparation of this chapter. In addition, further

---

48   Hannah Lewi and Stephen Neille, op. cit.
49   Ibid., 21.
50   Ibid., 23.

thanks are due to NTAV colleagues: Carmel Ganino, Ann Gibson, Judy Mahoney and Martin Purslow (Chief Executive Officer) and Neryl Stacey. I am grateful to the street artists who spoke with me about their perspectives.

Bibliography

'Paint-can curse gets historic daubing', *Melbourne Times*, April 21 (1999): 3.

Australia ICOMOS, *The Burra Charter: The Australia ICOMOS Charter for Places of Cultural Significance*, Victoria: Australia ICOMOS, 1999.

Burchall, G. 'Artist's legacy fades but not forgotten', *The Age*, June 12, 2008. http://www.theage.com.au/news/film/bfilmb-artist-keith-harings-legacy-fades-but-is-not-forgotten/2008/06/11/1212863732691.html (accessed November 3, 2008).

Brown, R. 'Melbourne graffiti considered for heritage protection', *The World Today*, ABC Radio, June 23 (2008): http://202.6.74.88/news/stories/2008/06/23/2282978.htm?site=melbourne.

Chalfant, H .and Prigoff, J. *Spraycan Art*. London/New York: Thames and Hudson, 1987.

City of Melbourne, Graffiti Management Plan, Melbourne, 2006: http://www.melbourne.vic.gov.au/rsrc/PDFs/Graffiti/graffitti_management_plan_2006.pdf.

City of Melbourne, *Registered Street Art Applications*, http://www.melbourne.vic.gov.au/info.cfm?top=145&pa=3274&pg=3842.

City of Melbourne, *Future Melbourne*, 2008: http://www.futuremelbourne.com.au/wiki/view/FMPlan/S1dGrowthAndStrategicAreas.

Cooper, M. and Chalfant, H. *Subway Art*. London/New York: Thames and Hudson, 1984.

Fannin, P. 'Slogan's Heroes – trust fights for history writ large', *The Age*, April 12 (1999): 3.

Houston, C. 'Anti-graffiti lobby sees red at heritage-listing proposal', *The Age*, June 23 (2008): 5.

Johnston, C. 'Social values – overview and issues', in *People's Places: Identifying and Assessing Social Value for Communities*. Canberra: Australian Heritage Commission, 1994a.

Johnston, C. *What is Social Value?* Canberra: Australian Government Printing Service, 1994b.

Lewi, H. and Neille, S. 'Drawing in and fading out: The case of the Commonwealth Games housing village, Western Australia', *Journal of Architectural Education* (2007): 15–21.

Masterson, A. 'Off the wall art', *The Age*, December 27, *Summer Age* supplement, (1994): 4–5.

Masterson, A. '…And so it was written', *The Age*, April 16 (1999): 16.

The National Trust of Australia (Victoria), The Cultural Heritage Centre for Asia and the Pacific at Deakin University, and Australia ICOMOS, *Intangible Cultural Heritage: A New Field of Endeavour*, Melbourne, 1–3 July 2008, unpublished conference proceedings.

National Trust of Australia (Victoria), *Classification Policy*. National Trust of Australia (Victoria): 2007.

National Trust of Australia (Victoria), Draft Classification Report for 'Keon Graffiti', Tanner Street, Richmond, File number 6969.

National Trust of Australia (Victoria), Minutes of Twentieth Century Buildings Committee, item 6 (h), Temporary file 'Keon Graffiti', Tanner Street, Richmond, April 8, 1998, 2.

Pascoe, R. 'Sport: The unbearable lightness of a bag of wind', in *Intangible Cultural Heritage: A New Field of Endeavour*, Melbourne, Australia, July 1–3, 2008, unpublished conference proceedings.

Robinson, D. *SoHo Walls: Beyond Graffiti*. London: Thames and Hudson, 1990.

State Government of Victoria, *Heritage Act 1995 (no. 93)*, Victoria: State Government of Victoria, 2007.

Topsfield, J. 'Brumby slams Tourism Victoria over graffiti promotion'. *The Age*, October 1 (2008): http://www.theage.com.au/articles/2008/10/01/122265114 0951.html.

UNESCO, *Convention for the Safeguarding of the Intangible Cultural Heritage*, Paris: UNESCO, 2003, http://unesdoc.unesco.org/images/0013/001325/ 132540e.pdf.

PART III
The Heritage of Housing

Chapter 8

The Georgian House:
The Making of a Heritage Icon

Peter Borsay

In 1995 there appeared *The Georgian Group Book of the Georgian House*. Profusely and beautifully illustrated, it could be dismissed as just another coffee-table book celebrating a glorious era of British architecture. However, though in all probability the coffee table was the location most copies were destined to adorn, the appearance of the book marked much more than that. It was a statement – refined and understated, implicit rather than explicit, but a statement nonetheless – of how far the Georgian house had travelled in its journey to becoming a heritage icon, and how this had and was being achieved. First, its contents were endorsed by not one but two members of the Royal Family, with a Foreword by the Group's patron the Queen Mother and an introduction by the Duke of Gloucester. Second, it was *The Georgian Group Book*, highlighting the perceived role of this organization, founded in 1937, in promoting and protecting the Georgian house. Third, in between its covers was not an endless parade of country house facades and interiors, but – after some brief introductory chapters – a practical analysis of the detailed features of Georgian house design; windows, doors, paint colour, wallpaper, lighting and such like. In other words, this was meant to be a practical book, not a glossy picture album, sophisticated not superficial heritage, for those discerning owners of a Georgian house who took their conservation responsibilities seriously. However, less this hinted at cultural snobbery and elitism, fourth, the cover blurb indicated that it was the ordinary man who was the intended target – that the Georgian house had a sort of 'mass' appeal – the book having 'been written with the owner or would-be owner of a modest family house or Georgian cottage primarily in mind'. Finally, though it was made clear that 'Today Georgian buildings are universally revered and admired' and 'everyone loves a Georgian house', this had not always been the case. The Duke of Gloucester [Richard, 1944–] reminded readers that 'it is hard to believe that my father's [Henry, 1900–1974] generation was brought up to believe that Georgian architecture was dull and repetitive'.[1] Meanwhile the author of the book declared that 'while there has always been a large degree of public affection [which clearly did not spread to the Duke of Gloucester's father or his generation] for Georgian buildings, this has

---

1 Steven Parissien, *The Georgian Group Book of the Georgian House*, (London: Aurum Press, 1995), 7, 11, 14.

not always been reflected in public policy', going on to cite the way in the late Victorian and inter-war years Georgian churches, terraces and squares in London were 'decimated', 'mutilated', and 'systematically destroyed', and how in the 1950s and 1960s 'architectural atrocities continued unabated'.[2]

Whether such cultural devastation could be attributed simply to negligent governmental policy, without wider public support, is a matter for debate; but there can be little doubt that attitudes to Georgian architecture have been transformed over time. In a nutshell, the heritage icon of the late twentieth century had to be *made*. The following essay is concerned with how and why this process occurred. It will explore the changing image and status of the Georgian house across five periods; post-Georgian/Victorian (1830–69), late Victorian and Edwardian (1870–1914), war and inter-war (1914–45), modern (1945–69) and post-modern (1970–).[3] The picture that emerges is of a highly mutable phenomenon, whose meaning regularly changed under pressure from wider economic, social and political forces. This raises fundamental questions about the way in which heritage in general, and the historic environment in particular, is valued. The examination of the Georgian house, and its elevation to the status of a heritage icon, suggests that value is historically negotiated – hence the need to study the history of heritage – and determined by factors largely extraneous to the object itself.

Post-Georgian and Victorian

The tidy boundary now drawn in works of history and the arts between the Georgian and Victorian periods would not have made much sense to those who occupied the immediate post-Georgian era. This was not only because the notion of Victorianism had not yet had chance to take shape, but also because the idea of something called 'Georgian', never mind a 'Georgian heritage', could at best have taken only the vaguest form. For those living in the immediate post-Georgian world their notion of what was modern and contemporary would still be defined by what was fashionable in the later Georgian period. That meant that what we now call the Georgian house continued to spread into the urban and rural landscape after 1830, replacing vernacular structures with brick boxes, and adding squares and long terraces of uniform housing that catered to the needs of the middle and skilled working class and echoed, albeit distantly, the fashionable buildings of eighteenth-century London, Bath, Edinburgh and Dublin.[4] However, looking back

---

2   Parissien, *Georgian Group Book*, 14–15.

3   This chronological framework is based on that developed in Peter Borsay, *The Image of Georgian Bath, 1700–2000: Towns, Heritage, and History*, (Oxford: Oxford University Press, 2000), 86–96, 143–206.

4   Peter Guillery, *The Small House in Eighteenth-Century London: A Social and Architectural History*, (New Haven and London: Yale University Press, 2004), 279–96; Roger Dixon and Stefan Muthesius, *Victorian Architecture*, (New York and Toronto:

to the late Georgian era also opened the door for the paradigm-shift in building design that we now associate with Victorianism, and which came to undermine the credentials of the Georgian house. At the heart of this was the breakdown in the dominance of classicism, and in particular of Palladianism. This was already evident by the mid-eighteenth century, which saw an opening up of stylistic opportunities and the first serious experiments in gothic revivalism, which later matured into a full blown 'gothic backlash'.[5] What characterized Victorianism, however, was not a rejection of classicism *per se*, since it remained a powerful strand in design, but the emergence of a thorough going stylistic eclecticism and historicism. Gothicism, and mutations of this such as Tudorbethan, were elements in this wide-ranging package. Where they held the edge over classicism, and in so doing undermined the status of the classical house, was in three key respects. First, they dovetailed nicely with the notion of what Mark Girouard has called the 'moral house'.[6] The Victorians considered themselves to be far in advance of their predecessors in matters of morality and religion. In these respects the eighteenth century was perceived by the nineteenth as culpably defective, fostering selfishness, corruption, loose personal standards in areas such as sexual relations and gambling, and irreligion.[7] The classical style was tainted by these associations. Gothicism, on the other hand, represented a re-engagement with an era of spiritualized Christian belief, symbolized by the architecture of the great cathedrals and monasteries. Gothic became *de rigueur* for new Anglican churches, but it could also be used to infuse the house, and the families it accommodated, with a new sense of moral purpose. Second, gothic acquired a particular appeal among members of the landed order, and those eager to join their ranks, keen to place space between themselves and the modern world of business. Adopting medieval styles and forms helped to insulate house owners from commerce and industry, launder any of their wealth that derived from these sources, and legitimate their authority by association with a distant feudal order. When the third Marquess of Bute, in alliance with the architect William Burgess transformed his Cardiff residence into the fantasy castle *par excellence* (1868–85), it was a way of insulating himself from the rising tide

---

Oxford University Press, 1978), 59–65; Donald J. Olsen, *The Growth of Victorian London*, (Harmonsdsworth: Penguin Books, 1979), 73–4, 77–8, 129–89; Mark Girouard, *The English Town*, (New Haven and London: Yale University Press, 1990), 185–8, 257–66; Barrie Trinder, *Beyond the Bridges: The Suburbs of Shrewsbury 1760–1960*, (Chichester: Phillimore, 2006); Christopher Train, 'The growth of Victorian Ludlow', in *Victorian Ludlow*, edited by David Lloyd et al., (Bucknell: Scenesetters, 2004), 32–52.

   5   Kenneth Clark, *The Gothic Revival; an Essay in the History of Taste*, (Harmondsworth: Penguin Books, 1964); Boyd Hilton, *A Mad, Bad and Dangerous People? England 1783–1846*, (Oxford: Clarendon Press, 2006), 475–87.

   6   Mark Girouard, *Life in the English Country House*, (Harmondsworth: Penguin Books, 1980), 267–98.

   7   Vic Gattrell, *City of Laughter: Sex and Satire in Eighteenth-Century London*, (London: Atlantic Books, 2006), 574–95; B.W. Young, *The Victorian Eighteenth Century: An Intellectual History*, (Oxford: Oxford University Press, 2007).

of urban expansion that lapped against his home, and of cleansing the profits from coal and commerce which financed the spectacular building programme.[8]

Third, gothic and its derivatives had suffered from one association so negative as to virtually proscribe its use for two centuries. Since the Reformation it was tarred with the brush of Catholicism and by implication anti-patriotism.[9] The extensive and decaying ruins of monasteries and abbeys stood as a permanent reminder of the spiritual degeneration of England before it was saved by Protestantism. It was of course perfectly obvious that from the Renaissance onwards Catholic countries also turned towards classicism, but this could not exorcize the English suspicion of the gothic style. Arguably it was only after the failure of the second Jacobite rebellion in 1745, and the removal of any serious threat of a restored Stuart and Catholic monarchy, that it became possible to reintroduce Gothicism into the palate of acceptable styles. That was to be a slow process, with the early gothic revival taking a more ornamental than structural form, but already before 1820 new country houses were being constructed in the style of churches, such as Fonthill Abbey (Wiltshire, begun 1796), and castles, such as Eastnor (Herefordshire, 1812–20). At this time the kudos of surviving pre-classical houses began to climb with the first wave of popular antiquarianism and country house visiting, focused primarily on sub-gothic Tudor and Jacobean mansions. The architecture of the 'olden time' and merry England now came to represent the essence of national identity.[10] By the start of Victoria's reign it is likely that less than half of new or rebuilt country houses were being constructed in the classical style, a proportion that fell to under one fifth by the early 1860s.[11]

The dissolution of the classical hegemony, and with it the influence of what has become known as the Georgian house, was driven by a complex but dynamic amalgam of three further factors; the spirit of improvement (the dominant motif of the age), a thirst for the past (that could both offset the impact of modernization at the same time as being progressive), and the generational syndrome, by which one generation defines itself by rejecting the norms of the previous.[12] None of

---

8   John Davies, *Cardiff and the Marquess of Bute*, (Cardiff: University of Wales Press, 1981); Martin Daunton, *Coal Metropolis: Cardiff 1870–1914*, (Leicester: Leicester University Press, 1977); Mark Girouard, *The Victorian Country House*, (Oxford: Clarendon Press, 1971), 123–50; John Newman, *The Buildings of Wales: Glamorgan*, (London: Penguin Books, 1995), 194–211.

9   Graham Parry, *The Trophies of Time: English Antiquarians of the Seventeenth Century*, (Oxford: Oxford University Press, 1995), 227–32; Rosemary Sweet, *Antiquaries: the Discovery of the Past in Eighteenth-Century Britain*, (London and New York: Hambledon and London, 2004), 238–47.

10   Peter Mandler, *The Fall and Rise of the Stately Home*, (New Haven and London: Yale University Press, 1997), 21–106.

11   Girouard, *Victorian Country House*, 5–6, 33.

12   Asa Briggs, *The Age of Improvement 1783–1867*, (London: Longman, 1959); Sarah Tarlow, *The Archaeology of Improvement in Britain, 1750–1850*, (Cambridge: Cambridge University Press, 2007); Charles Dellheim, *The Face of the Past: The Preservation of the*

these were unique to the post-Georgians, but in the hands of the wealth-expanding and self-confident society that constituted Victorian Britain the potent mixture of factors provided the necessary thrust to break with the Georgian past.

Late Victorian and Edwardian

That thrust, and the self assurance and unprecedented prosperity that underpinned it, was beginning to wane a little in the years before the First World War. It was not that national wealth and power ceased to grow; indeed, working class real incomes rose substantially, and the Empire grew rapidly to reach its territorial zenith in 1921. However, though the so-called 'Great Depression' of the late nineteenth century was in many respects an economic myth, it did possess an element of psychological reality, in that Britain was no longer able to enjoy the considerable economic benefits of being the first industrial nation as other countries, such as Germany and the USA, rapidly industrialized and emerged as powerful competitors, pegging back profit margins. This prompted in Britain a degree of cultural self-reflection and consolidation, which manifested itself in a surge of nostalgia, 'pastoralism' and anxiety for what had and was being lost though urbanization and industrialization. With this came the establishment of a series of legislative initiatives and institutions geared to protecting and preserving the national past, such as the Society for the Protection of Ancient Buildings (1877), the *Ancient Monuments Acts* (1882, 1900, 1913), the National Footpaths Preservation Society (1884), the Survey of London (1894), the National Trust (1895, *National Trust Act* 1907), *Country Life* (1897) and the Royal Commission on Historical Monuments (1908).[13] It was at this point in time that the modern heritage movement began to take shape. With a decent chronological gap between the Georgian and late Victorian eras, and the Georgian houses's association with a pre-industrial society, it was theoretically in a strong position to take advantage of the new climate. In practice its capacity to do this was severely limited by a number of factors. First, architecture of the 'olden time' still represented the dominant historic fashion. The early volumes of *Country Life* were filled with articles and photographs of

---

Medieval Inheritance in Victorian England, (Cambridge: Cambridge University Press, 1982); Olsen, Victorian London, 55.

    13   G.R. Searle, *A New England? Peace and War 1886–1918,* (Oxford: Clarendon Press, 2004), 172–86, 602–13; Michael Hunter, 'Introduction: The fitful rise of British preservation', and Chris Miele, 'The first conservation militants: William Morris and the Society for the Protection of Ancient Buildings', in *Preserving the Past: The Rise of Heritage in Modern Britain,* edited by Michael Hunter, (Stroud: Alan Sutton, 1996), 6–10, 17–56; John Delafons, *Politics and Preservation: A Policy History of the Built Heritage, 1882–1996,* (London: E & FN Spon, 1997), 16–35.

restored medieval castles and manor houses.[14] However, the parameters of what constituted historic style had broadened considerably compared with earlier in the century. William Morris and the Arts and Crafts movement were drawing into the net a widening range of styles that could be defined in some way as hand-crafted and vernacular. This opened the door to domestic classicism. New Arts and Crafts houses made plentiful use of hung tiles, bricks and gabled dormers, and the so-called Queen Anne style flourished on the streets of late Victorian and Edwardian towns and cities.[15]

But there were limits as to how far classicism could be accommodated, and it was here that it hit up against the second major barrier to its acceptance; national sentiment. Architectural theory and practice in the period was underpinned by an imperial, nationalist and racist agenda.[16] Classicism was at heart an international style, with universal rules. As such it would not easily appeal to the contemporary *zeitgeist*. However, the negative implications of this could be neutralized so long as the specific form of classicism promoted was seen to reflect domestic roots and incorporate local idiosyncrasies. The *Georgian* house fell foul of this by virtue of its version of classicism and its chronology. 1715 (with the publication of the first volumes of Colen Campbell's *Vitruvius Britannicus* and Nicholas Dubois' English translation of Palladio's *Four Books*) is the date traditionally taken to mark the introduction of Palladianism in Britain, a style which deliberately drew classicism back to its Italian roots, cleansing it of many of its local anomalies, a point confirmed by the xenophobic reaction of contemporaries like Hogarth who criticized it as anti-English. A year earlier, 1714, was also the date of the accession of George I and the establishment of the Hanoverian dynasty in England. Given its Germanic origins there was the feeling, at the time and subsequently, that there was something un-English about the new regime, and the culture which grew up around it. In the late nineteenth century, with the emergence of a unified Germany, and the intense economic and political competition between Britain and Germany that preceded World War I, these resonances took on a particular significance. The Royal Commission on Historical Monuments, founded in 1908, used as its original terminal point of record 1700; when in 1921 this was revised it was to advance it no further than 1714. The upshot was that though the vernacular and the mildly 'baroque' versions of classicism that flourished in the later Stuart

---

14   John Cornforth, *The Inspiration of the Past: Country House Taste in the Twentieth Century*, (Harmondsworth: Viking, 1985), 20–46; John Cornforth, *The Search for a Style: Country Life and Architecture, 1897–1935*, (New York and London: W.W. Norton and Company, 1989), 57; Roy Strong, *Country Life 1897–1979: The English Arcadia*, (London: Country Life Books and Boxtree, 1996), 54–9.

15   Elizabeth Cumming and Wendy Kaplan, *The Arts and Crafts Movement*, (London: Thames and Hudson, 1991), 55; Mark Girouard, *Sweetness and Light: The Queen Anne Movement, 1860–1900*, (Oxford: Clarendon Press, 1977).

16   David Watkin, *The Rise of Architectural History*, (London: Architectural Press, 1983), 94–115.

period was surveyed and praised (and Christopher Wren [1632–1723] and the 'Wrenaissance' celebrated), Georgian classicism (which in practice was by no means wholly Palladian and had little to do with Germany) received much less recognition and fewer plaudits. In his influential text on the subject of classicism, *History of Renaissance Architecture in England, 1500–1800* (1897), Sir Reginald Blomfield could write with scarcely disguised disdain of 'the fine, if somewhat frigid, architecture of the first half of the eighteenth century'.[17]

If the Georgian house struggled to realize its potential during this period there was evidence, nonetheless, that it had started on the long road to becoming a heritage icon. It was not just that the general context was improving with the nostalgic turn, the growth in domestic tourism and the rise of the heritage movement; there were also the first signs that Georgian architecture and the Georgian house were beginning to be understood and valued. Some of the first intimations of this were to be seen in London and Bath, an unsurprising state of affairs given the wealth of Georgian fabric still extant in these places.[18] One reason for its survival in Bath was the fact that Victorian property development had been focused in the suburbs of the city, leaving the Georgian core neglected but relatively intact. However, by the late nineteenth century schemes were afoot, in the interests of improvement generally and in particular to upgrade the spa facilities, to modernize parts of the city centre. Plans were drawn up to extend the Grand Pump Room Hotel which involved the demolition of the north side of the colonnaded Bath Street, an elegant piece of late eighteenth-century town planning (Thomas Baldwin, 1791–94) that linked the water facilities. A campaign was mounted to resist the changes, the Old Bath Preservation Society was founded in 1909 to lead the fight, and the first battle to save a piece of the spa's Georgian landscape was won. This was possible because of three local factors: a late Victorian and Edwardian rise of interest in and nostalgia for the spa's history, which culminated in a the spectacular pageant of 1909, and which had also seen the erection of wall plaques from 1899 on the houses of celebrated (largely Georgian) visitors and residents; a growing awareness of the tourist potential of the city; and the first serious survey and study of the city's Georgian buildings, published by the Bath architect Mowbray Green in 1904.[19] The Bath Street victory was an important harbinger of things to come. But it was no more than a sign. Bath's Georgian buildings were valued primarily for their personal associations rather than their architectural merits. Moreover, it is clear that there was precious little appreciation or awareness, even among the architectural profession, of the Georgian jewels preserved in Bath, never mind other provincial towns. After Mowbray Green had presented a paper on the 'The Eighteenth-century Architecture of Bath' to a meeting of the Architectural

---

17   Reginald Blomfield, *A History of Renaissance Architecture in England*, 2 vols. (London: George Bell, 1897), II, 213.

18   Olsen, *Victorian London*, 85–6.

19   Mowbray Green, *The Eighteenth Century Architecture of Bath* (Bath; George Gregory, 1904); Borsay, *Image of Georgian Bath*, 72–3, 150–7.

**Figure 8.1   The north colonnade of Bath Street, Bath (architect, Thomas
            Baldwin, 1791–94). The site of the successful battle in 1909 to
            save a piece of the city's Georgian townscape**
Source: © Peter Borsay

Association in 1901, the member proposing the vote of thanks observed that
'although he had been to Bath he had no idea that there was so much of interest
there'.[20]

---

20   Mowbray Green, 'The Eighteenth Century Architecture of Bath', *Builder* (27
April 1901): 412.

War and Inter-War

In many respects the two world wars, and the years framed by them, were highly unpropitious ones for the Georgian house. As campaigners at the time and subsequently have made clear, this was a period of major losses among Georgian town houses; during the 1920s and 1930s in London 'many fine Georgian townscapes – most notably Georgian Bloomsbury, Adam's Adelphi and Nash's Regent Street – were systematically destroyed', and some of the capital's finest squares and town houses demolished or altered beyond recognition.[21] Swathes of sub-Georgian artisan and working class urban terracing must also have been swept away as the inner city slum clearance schemes introduced in the later nineteenth century were extended. This was also a period when the country house, of which Georgian houses formed a large number was under enormous pressure. A combination of agricultural depression and Government fiscal policy was encouraging, and in many cases forcing the landed aristocracy and gentry to sell up their estates where they could, and where they could not, close up, dismantle, or simply fail to maintain their great houses so that ruination and demolition became inevitable. This was accompanied, as Peter Mandler has argued, by 'the general public's near total indifference to the fate of the country house in the 1920s and 1930s'.[22]

Yet the situation for the Georgian house was not entirely bleak. First, the nostalgia and passion for the pastoral which captured the sensibilities of the late Victorian and Edwardian middle class if anything intensified in the inter-war years, fuelled by the cult of fitness, increasing opportunities for mobility and a plethora of tourist and topographic publications aimed at revealing hidden Britain.[23] Admittedly, much of this was directed at the countryside and quaint cottages, but all suitably historic structures benefited to some degree. Second, among the political and cultural elite, and some within the architectural profession, Georgian taste began to acquire an aesthetic kudos and chic status. In part this was the effect of the generational syndrome, that a century previously had done so much to undermine classicism. Now the target was Victorianism and high Gothicism with all its pompous moral and architectural pretensions and its sheer aesthetic fussiness. The most tangible effect of this elite reaction was the establishment of the Georgian Group in 1937.[24] Third, generational antipathy was instigating a potentially even more violent reaction in the architectural profession, with the emergence of the Modern Movement. The Georgian house hardly constituted

---

21   Parissien, *Georgian Group Book*, 14–15.

22   Mandler, *Stately Home*, 154.

23   Sean O'Connell, *The Car and British Society: Class, Gender and Motoring 1896–1939*, (Manchester: Manchester University Press, 1998); David Matless, *Landscape and Englishness*, (London: Reaktion, 1998), 62–100.

24   Cornforth, *Inspiration of the Past*, 47–84; Watkin, *Rise of Architectural History*, 115–34; Gavin Stamp, 'Origins of the Group', *Architectural Journal*, 175: 13 (1982), 35–8.

'the shock of the new', but its clean lines and functionalism in comparison to the Victorian house raised its profile among disciples of modernism.

Fourth, a group critical in shaping the pattern of architecture in the twentieth century, who in the early years of the century were forging and disseminating their agenda and establishing themselves as a profession within a profession, were the town planners. Modernism offered one source of inspiration, but up until the Second World War equally if not more influential was classicism, whose associations with urbanism, and in particular with Classical and Renaissance town planning, provided a rich body of theory and practice to draw upon.[25] The internationalism of classicism would also have probably appealed to most architects as a release from the disease of nationalism that had infected traditional architectural taste. However, if local inspiration was required, and an injection of patriotism needed, eighteenth-century Britain provided some valuable models, not least in London, Edinburgh, Dublin and Bath. In the last of these the idea of the model planned city now figured frequently in interpretations of its Georgian architecture, and during the inter-war years the spa became something of a Mecca for architects and planners. RIBA held the British Architects Conference there in 1928, and the Chartered Surveyors' Institution its annual country meeting in 1933. On the latter occasion delegates were informed that the city 'has an exceptional interest to a surveyor from the point of view of town planning', and that the parts of the town developed by the Woods (John I 1704–54, John II 1728–81) 'remains one of the earliest and best examples of town-planning in the country'.[26] On the back of planning for change also came planning for conservation, with the first indications, beyond the limited protection of the *Ancient Monuments Act*, of what was to emerge as the listing process.[27] Finally, the foundation of the Georgian Group in 1937 was a sign not only that architectural taste was changing, but also that those with influence and power were willing to *organize* themselves to protect what survived; the Group was established because of the immediate threat to demolish Nash's Carlton House Terrace in London. The Georgian house, in other words, had become worth fighting for.

Post-War

Aerial bombing during the Second World War placed many city buildings under threat and there were inevitable losses of historic fabric. From the heritage perspective particularly pernicious were the so-called *Baedeker* raids in which historic British cities were deliberately targeted in reprisal for the allied bombing

25   Helen Meller, *Towns, Plans and Society in Modern Britain*, (Cambridge: Cambridge University Press, 1997), 25–65; T. Harold Hughes and E.A.G. Lamborn, *Towns and Town-planning: Ancient and Modern*, (Oxford: Clarendon Press, 1923).

26   Chartered Surveyors' Institution. Annual Country Meeting, Bath, 1933 (Bath), 3, 17.

27   Andrew Saint, 'How Listing Happened', *Preserving the Past*, 115–20.

of the ancient Hanseatic port of Lübeck. York, Canterbury, Norwich, Exeter and Bath were the victims, with the eighteenth-century architecture in the last two suffering particular damage.[28] If the Georgian house had a bad war, it is arguable that it had an even worse peace. Propaganda during the war had exploited ideas of heritage to strengthen national identity and unity, and there was predictable outrage at the targeting of historic cities and monuments during the *Baedeker* bombings; but, once the fight was over, far from heightening a sense of the past the war seems to have engendered a thirst for wholesale change. This opened the door for the town planners and advocates of modernism, but is should be stressed that their interventionist and architecturally innovatory agenda had an appeal not limited to the professionals. During the 1950s and 1960s substantial chunks of generally minor Georgian town housing were lost as city and town centres were redeveloped along modernist lines to provide more comfortable, efficient and democratic spaces in which to accommodate the mass of citizens and to transact their economic and social business. Part of this meant constructing an efficient transport infrastructure that could service the swelling number of motor cars, later to be portrayed as the villain of the peace in damaging the historic built environment. Outside the towns matters seemed scarcely any better; the 1950s and 1960s were quantitatively by far the peak decades for country house demolition for the entire century after 1875.[29]

Despite the patent enthusiasm for change and modernity, and the destruction of many eighteenth-century properties, the position of and prospects for the Georgian house were far from entirely bleak. The gains made earlier in the century were not lost. Architects and town planners, though they now embraced modernism on a wide scale, did not turn against the Georgian house *per se*. There was still seen to be an affinity between modernism and classicism, and there was intellectual ammunition increasingly available to woo the professionals. Serious works of scholarship, cataloguing and analysing the Georgian contribution to the townscape were now beginning to appear in numbers, several by architects themselves. No major study of Bath's eighteenth-century architecture had appeared since that by Mowbray Green in 1904. Walter Ison's *Georgian Buildings of Bath* (1948) meticulously recorded in text and photographs the city's eighteenth-century heritage, and whereas Green had focused heavily on the era of the Woods the new study took full cognizance of the late Georgian contribution. Ison followed up this with a parallel study of Bristol, which appeared in 1952 the same year as Maurice Craig's path-breaking volume on *Dublin 1660–1860*. However, perhaps the most influential work to appear on the eighteenth-century townscape was John Summerson's scholarly but brilliantly accessible *Georgian London*, which originated in lectures and articles prepared before the war, but was only published in 1945. It may be that one factor behind Summerson's appeal was that he was

---

28   Niall Rothnie, *The Baedeker Blitz: Hitler's Attack on Britain's Historic Cities*, (Shepperton: Ian Allan, 1992).
29   Mandler, *Stately Home*, 360.

both an advocate of Georgian design and the Modern Movement.[30] Many of these writers explored and extolled the Georgian town from a planning perspective,[31] and the town planners repaid the compliment by a paradoxical mixture of destruction and conservation. On the one hand, the vision and actions of figures like the Woods (who had little compunction at the time in 'improving' or removing historic fabric) in creating a progressive and planned landscape in the eighteenth century justified and required similar intervention in post-war Britain; on the other hand, respect for the architectural heritage (not least that of the Georgians) created a responsibility to preserve, as long as this did not unduly block progress, at least the major features of the historic built environment. These attempts to service the requirements of both progress and preservation, to reconcile – as John Pendlebury has put it – 'history and modernity', were reflected in the ambitious post-war plans of figures such as Patrick Abercrombie and Thomas Sharp for individual historic cities, many with strong Georgian elements like Bath, Warwick and Durham.[32] It was also out of this context that the listing and planning regime emerged, embodied in the *Town and Country Planning Acts* of 1944 and 1947 that established the formal framework for protecting historic domestic architecture.[33]

The charge that Georgian (post-1714) architecture was un-English had by the Second World War largely evaporated. In 1946 the Royal Commission on Historical Monuments was permitted to extend its brief to 1850. However, it was one thing to remove the negative, quite another to assert the positive; that is, to claim that the Georgian house in some way embodied national identity. Matters were not helped by the general public indifference to the fate of gentry and aristocratic landowners and their houses that had prevailed during the first half of the century. In post-war Britain it would seem that a new democratic national identity was being forged to

---

30   Walter Ison, *The Georgian Buildings of Bath*, (London: Faber and Faber, 1948); Walter Ison, *The Georgian Buildings of Bristol*, (London: Faber and Faber, 1952); Maurice Craig, *Dublin 1660–1860*, (London: Cresset Press, 1952); Elizabeth McKellar, 'Populism Versus Professionalism: John Summerson and the Twentieth-Century Creation of the "Georgian"', in *Articulating British Classicism: New Approaches to Eighteenth-Century Architecture*, edited by Barbara Arciszewska and Elizabeth McKellar, (Aldershot: Ashgate, 2004), 35–56.

31   See, for example, John Summerson, 'John Wood and the English Town-planning Tradition', in John Summerson, *Heavenly Mansions and Other Essays on Architecture*, (London: Cresset Press, 1949), 87–110.

32   Patrick Abercrombie, John Owens and H. Anthony Mealand, *A Plan for Bath*, (Pitman, 1945); Patrick Abercrombie and Richard Nickson, *Warwick: Its Preservation and Redevelopment*, (London: Architectural Press, 1949); Thomas Sharp, *Cathedral City: A Plan for Durham*, (London: Architectural Press, 1945); John Pendlebury, 'Planning the historic city: Reconstruction plans in the United Kingdom in the 1940s', *Town Planning Review*, 74, 4 (2003): 371–93; John Pendlebury, 'Reconciling history with modernity: 1940s Plans for Durham and Warwick', *Environment and Planning B: Planning and Design*, 31 (2004): 331–48.

33   Delafons, *Politics and Preservation*, 56–61; Saint, "How Listing Happened", 120–33.

Figure 8.2   Berrington Hall, Herefordshire (Henry Holland, c. 1778–81). Surrendered by Lord Cawley to the Treasury in 1957 in part payment of estate duty, and transferred to the National Trust

Source: © Peter Borsay

which the landowner and his Georgian house had little relevance. Yet moves were afoot which would change this. For much of its early life the National Trust was 'a small and exclusive society',[34] primarily devoted to holding and protecting chunks of the countryside, though buildings had always been within its remit. In the late 1930s efforts were made by a pressure group within the Trust to turn it into a vehicle for protecting the heritage of the country house and to do so by negotiating favourable inheritance tax arrangements with the Government. Only limited headway was made before the war, though the *National Trust Act* of 1937 established the framework for transferring properties to the Trust with an endowment and leaseback arrangement. However, the establishment of the National Land Fund in 1946 and taxation provisions in the *Finance Acts* of 1951 and 1953 set up the critical machinery for what was to prove in the long run the transmission of a sizeable body of houses, many of which came with their owners, into Trust ownership.[35] Though state nationalization was avoided cultural nationalization swept ahead. In one fell swoop the takeover by the

---

34   Mandler, *Stately Home*, 197.

35   Mandler, Stately Home, 265–353; Jennifer Jenkins and Patrick James, *From Acorn to Oak Tree: The Growth of the National Trust 1895–1994*, (London: Macmillan,

National Trust placed country houses at the centre of the national heritage. Many of these houses were Georgian, a profile strengthened by a continuing anti-Victorian aesthetic which focused alterations and demolitions on post-Georgian fabric. The change in image was one that surviving country houses were to benefit from whether in the ownership of the Trust or not. In the post-war decades a growing number of houses began to open to the public. Many of these were Trust properties, sustained by a climbing membership (7,000 in 1945, 150,000 by 1964); but the majority were in private ownership, as more favourable economic circumstances and a rising demand among the public to visit, encouraged a policy of commercial exploitation.[36] Ironically, this demand owed much to the enemy of the historic townscape, the motor car, whose numbers continued to escalate rapidly after the war (increasing three-fold merely between 1950 and 1962), providing just the sort of family-friendly flexible access required to reach the hidden corners of the countryside.[37]

Post-Modern

In the years after the Second World War the Georgian house secured its position as part of the national heritage. However, it was only over the latter decades of the twentieth century that it became a true heritage icon. In part this was simply a function of the fact that the whole heritage agenda, and the industry that supported it, geared up several notches under the impact of an amalgam of economic, social and political factors. Perhaps the single most dramatic sign of this heritagization of the nation was the soaring membership of the National Trust; in 1965 it stood at 158,000, by 1990 at just over 2 million, and by 2008 at 3.5 million (with 12 million people visiting pay for entry properties and an estimated 50 million open air sites).[38] No wonder in 1994 the British historian David Cannadine, then working in the USA, could lament 'as the twentieth century draws to a close, the image of Britain which is projected abroad (and at home) becomes even more that of a Ruritanian theme park, a contrived fantasy of hype and heritage', and two years earlier the archaeologist Peter Fowler speculated, 'I doubt if anyone before this moment can have been subject to quite so much "pastness" as there is now around *and be so aware of it*'.[39] Within the specific context of the built environment two

---

1994), 74–164; John Gaze, *From Acorn to Oak Tree: The Growth of the National Trust 1895–1994*, (London: Macmillan, 1988), 113–33, 145–57.

    36   Mandler, *Stately Home*, 355–400.

    37   Philip Bagwell and Peter Lyth, *Transport in Britain: From Canal Lock to Gridlock*, (London and New York: Hambledon and London, 2002), 113.

    38   Jenkins and James, *Acorn to Oak Tree*, 337; The National Trust, 'The Charity', http://www.nationaltrust.org.uk/main/w-trust/w-thecharity.htm, accessed 14.3.08.

    39   David Cannadine, *Aspects of Aristocracy: Grandeur and Decline in Modern Britain*, (London: Penguin Books, 1995), 242; Peter Fowler, preface to *The Past in Contemporary Society: Then, Now*, (London and New York: Routledge, 1992), xvii.

major changes occurred which raised the profile of historic fabric. First, there was a strong reaction against the Modern Movement and the dominant grip it achieved on the new built environment in the 1950s and 1960s. Architecturally speaking post-modernism did not mean the end of modernism, which continued to have its advocates; but it did entail a much greater engagement with historic styles. This led in a minority of cases to a self-consciously ironic mixture of old and new; in an even smaller number of cases, to authentic reproduction of period houses; and in the majority of cases to a rifling of the pockets of the past for individual elements – a pilaster here a pediment there – appended to a box for living, as in the homes erected on thousands of new housing estates. Second, the planning agenda, which had been carried forward on the wave of enthusiasm for the new Britain of the Welfare State, began to unravel, as the post-war political consensus and corporatism dissolved, conservatism (with a small and capital C) was re-energized, and the primacy of private property was re-established. Conservation rather than demolition and rebuild came to play an increasingly important role in the planning process, a reorganization of priorities driven by a more organized and aggressive approach among the amenity organizations (such as the Georgian Group), and a string of high profile campaigns to save parts of the historic landscape under threat.[40]

These forces were in large measure common to the reception of the past as a whole; the Victorian (together with the Arts and Crafts and Art Deco) house began to climb back into fashion soon after the war – there were signs of a revival in Victorian studies as early as the 1930s[41] – and before the end of the Millennium even 1950s and 1960s properties were being drawn into the heritage net. Such an eclectic approach to the past risked spreading the jam too thinly. Rival periods can reinforce each other – the public may see the past all as one amorphous product – and it is arguable that there was so much demand around for pastness in the period that there was space for every contender at the feeding table. But the past is a competitive business, and one era's gain is potentially another's loss. How did the Georgian house fare? In the case of the country house, with the long eighteenth century figuring so prominently in the National Trust's corpus of properties and of other houses open to the public, the answer would seem to be, very strongly. The Georgian country house benefited from the long-term relocation of the *idea* of England to the countryside that gathered pace from the late nineteenth century, and the portrayal of the Georgian era itself as the Indian summer of rural England before industrialization and urbanization transformed the nation. As such the Georgian country house was in a strong position to claim iconic status. Whether the Georgian *town* house could draw on such sentiments and acquire a similar

40   Delafons, *Politics and Preservation*, 133–88; Peter J. Larkham, *Conservation and the City*, (London and New York: Routledge, 1996).

41   Martin Hewitt, 'Culture or society? Victorian studies, 1951–64', in *The Victorians since 1901: Histories, Revisions and Representations*, edited by Miles Taylor and Michael Wolff, (Manchester: Manchester University Press, 2004), 90–104.

Figure 8.3   27 Broad Street, Ludlow
Source: © Peter Borsay

kudos was less clear, particularly given the 'National Trust's huge portfolio of country properties compared to its tiny portfolio of town ones'.[42] Matters were not helped by the manner in which before and after the Second World War Georgian urban classicism had been so closely associated with the increasingly discredited town planning movement.

Nonetheless, the Georgian town house also rose to iconic status alongside its country partner. How has this happened? First, though rural nostalgia and anti-urbanism were undoubtedly features of British culture in the twentieth century, the majority of people continued to live in towns and cities. This was not simply a question of necessity. For all its faults, the urban environment held considerable advantages over the countryside when it came to access to work and educational, health, consumer, recreational and cultural services. The fashionable re-colonization of city and town centres in the later twentieth century provided strong evidence of this, a process reinforced by the changing pattern of family life and age profile of society. Moreover, towns and areas within towns vary enormously from each other. The image of urban blight associated with the industrial cities did not adhere so strongly to the historic towns. Often well endowed with Georgian fabric (especially so in the case of the national capitals, county towns, spas and some

---

42   Girouard, *English Town*, 313.

seaside resorts), they became highly desirable places to reside in. This was also the case for the multiplicity of smaller country towns (such as Ludlow, Stamford and Totnes) which, combining the benefits of country and urban living, saw something of a renaissance. Second, the Georgian town house enjoyed an almost unblemished positive press. From basic tourist literature to more sophisticated texts a wealth of material – drawing upon a continuing discourse established in the inter-war years of the eighteenth-century town house as the model of civilized living – was published to whet the public's appetite for and raise its understanding of urban classicism.[43] This was reinforced by a general Georgamania (fuelled by TV and film adaptations of Jane Austen novels, in which towns like Bath played a prominent role) and by a far more organized and pro-active Georgian Group, many of whose members lived in or close to London. Third, historic properties in favourable locations benefited from the long-term rise in urban property values from the 1970s. Even artisan dwellings which in the 1940s and 1950s could be dismissed as minor Georgian and seen as expendable in the interests of planning schemes, by the latter decades of the twentieth century – as property values begin to escalate – were seen as worth fighting bitter campaigns to save and were being lovingly restored in 'authentic' fashion. Moreover, the association of Georgian urban classicism with large-scale urban planning was quietly dropped, as either eighteenth-century planning was redefined as 'green' planning, or the focus was shifted from the grand set pieces to the individual house and its multiple parts. Fourth, underpinning all these changes was the rising wealth, numbers and influence of what has been called the 'service class', a group within the middle class whose managerial and professional skills were in high demand in a modernizing economy, and who were seeking the sophisticated cultural tools to establish their status.[44] The success of urban classicism in the eighteenth century had in large measure been built on its capacity to express complex messages of status for an expanding middling order, keen to differentiate itself from the common people. In the later twentieth century the potential for expressing status remained the same, though now it was directed more to servicing *intra-* rather than *inter-* class divisions.

Despite their image of timelessness heritage icons are as much playthings of fashion as shoes and motor cars. They have to be made, need regular re-modelling and can easily be forgotten. Heritage therefore has a history. For much of the nineteenth century the Georgian house was eclipsed by its Victorian successor. Only towards the very end of the century, during that critical period when the modern heritage industry began to emerge, were the first signs of recovery discernible in places like London and Bath. The founding of the exclusive Georgian Group in

---

43   See, for example, Dan Cruickshank and Neil Burton, *Life in the Georgian City*, (London: Viking, 1990); Girouard, *English Town*; Parissien, *Georgian Group Book*.

44   Nigel Thrift, 'The Geography of Late Twentieth-century Class Formation', in *Class and Space: the Making of Urban Society*, edited by Nigel Thrift and Peter Williams, (London: Croom Helm, 1987), 207–53; John Urry, *The Tourist Gaze: Leisure and Travel in Contemporary Societies*, (London: Sage, 1990), 88–90, 96–7, 113–14.

1937 was both a sign of the continuing rise in status of the Georgian house, but also a reaction to the indifference among public and Government alike towards the damage being sustained by Georgian properties. After the Second World War the modernist and planning agendas, and the continuing rise in country house demolitions, left the Georgian country and town house vulnerable. However, beneath the surface important changes were underway that laid the foundations for the dramatic climb to iconic status that was seen in the latter decades of the century. Royal patronage of the Georgian Group today is a sign both of the kudos enjoyed by the Georgian house, and of the way it has come to embody national identity. When the Queen Mother died she was replaced as patron of the Group by the second most senior member of the Royal Family, Prince Charles. This was entirely appropriate. He was a savage critic of 'the fashionable theories of the 50s and 60s … [which] have spawned deformed monsters which have come to haunt our towns and cities'. Moreover, he advocated, as his architectural 'manifesto', *A Vision of Britain* (1990) made clear, a return to the values and rules that underpinned eighteenth-century classicism, praising historic 'towns such as Cheltenham and Bath [which] exemplify the virtues of architectural harmony; not only in their layout but in their organization of the smaller architectural elements'.[45] To 'implement the principles expounded in' *The Vision* there began in the early 1990s the erection of the extraordinary settlement of Poundbury on the Prince's Duchy of Cornwall property adjacent to Dorchester. Built in a predominantly if not exclusively classical style, and intended (when completed in around 2025) to accommodate about 5,000 people, its construction demonstrates just how far the Georgian house has come as a heritage icon, and the social, cultural and commercial forces that underpinned this architectural apotheosis.[46]

Bibliography

Abercrombie, Patrick, John Owens, and H. Anthony Mealand. *A Plan for Bath.* Pitman, 1945.

Abercrombie, Patrick and Richard Nickson. *Warwick: Its Preservation and Redevelopment.* London: Architectural Press, 1994.

Bagwell, Philip, and Peter Lyth. *Transport in Britain: From Canal Lock to Gridlock.* London and New York: Hambledon and London, 2002.

---

45   Charles, Prince of Wales, *A Vision of Britain: A Personal View of Architecture*, (London: Doubleday, 1989), 7, 85.

46   The Duchy of Cornwall, Design & Development, Poundbury, http://www.duchyofcornwall.org/designanddevelopment_poundbury.htm, accessed 17.3.08; Peter Borsay, 'From Bath to Poundbury: The rise, fall and rise of polite urban space 1700–2000', in *Cities in the World, 1500–2000*, edited by Adrian Green and Roger Leech, (Leeds: Maney, 2006), 97–115.

Blomfield, Reginald. *A History of Renaissance Architecture in England*. 2 vols. London: George Bell, 1897.

Borsay, Peter. *The Image of Georgian Bath, 1700–2000: Towns, Heritage, and History*. Oxford: Oxford University Press, 2000.

Borsay, Peter, 'From Bath to Poundbury: The rise, fall and rise of polite urban space 1700–2000'. In *Cities in the World, 1500–2000*, edited by Adrian Green and Roger Leech, 97–115. Leeds: Maney, 2006.

Briggs, Asa. *The Age of Improvement 1783–1867*. London: Longman, 1959.

Cannadine, David. *Aspects of Aristocracy: Grandeur and Decline in Modern Britain*. London: Penguin Books, 1995.

Charles, Prince of Wales. *A Vision of Britain: A Personal View of Architecture*. London: Doubleday, 1989.

Chartered Surveyors' Institution. Annual Country Meeting, Bath, 1933. Bath.

Clark, Kenneth. *The Gothic Revival: An Essay in the History of Taste*. Harmondsworth: Penguin Books, 1964.

Cornforth, John. *The Inspiration of the Past: Country House Taste in the Twentieth Century*. Harmondsworth: Viking, 1985.

Cornforth, John. *The Search for a Style; Country Life and Architecture, 1897–1935*. New York and London: W.W. Norton and Company, 1989.

Craig, Maurice, *Dublin 1660–1860*. London: Cresset Press, 1952.

Cruickshank, Dan and Burton, Neil. *Life in the Georgian City*. London: Viking, 1990.

Cumming, Elizabeth, and Wendy Kaplan. *The Arts and Crafts Movement*. London: Thames and Hudson, 1991.

Daunton, Martin. *Coal Metropolis: Cardiff 1870–1914*. Leicester: Leicester University Press, 1977.

Davies, John. *Cardiff and the Marquess of Bute*. Cardiff: University of Wales Press, 1981.

Delafons, John. *Politics and Preservation: A Policy History of the Built Heritage, 1882–1996*. London: E & FN Spon, 1997.

Dellheim, Charles. *The Face of the Past: the Preservation of the Medieval Inheritance in Victorian England*. Cambridge: Cambridge University Press, 1982.

Dixon, Roger, and Stefan Muthesius. *Victorian Architecture*. New York and Toronto: Oxford University Press, 1978.

Fowler, Peter. *The Past in Contemporary Society: Then, Now*. London and New York: Routledge, 1992.

Gattrell, Vic. *City of Laughter: Sex and Satire in Eighteenth-century London*. London: Atlantic Books, 2006.

Gaze, John. *Figures in a Landscape: a History of the National Trust*. Barrie and Jenkins, 1988.

Girouard, Mark. *The Victorian Country House*. Oxford: Clarendon Press, 1971.

Girouard, Mark. *Sweetness and Light: The Queen Anne Movement, 1860–1900*. Oxford: Clarendon Press, 1977.

Girouard, Mark. *Life in the English Country House*. Harmondsworth: Penguin Books, 1980.

Girouard, Mark. *The English Town*. New Haven and London: Yale University Press, 1990.

Green, Mowbray. 'The Eighteenth Century Architecture of Bath', *Builder* (27 April 1901): 407–11.

Green, Mowbray. *The Eighteenth Century Architecture of Bath*. Bath: George Gregory, 1904.

Guillery, Peter. *The Small House in Eighteenth-Century London: A Social and Architectural History*. New Haven and London: Yale University Press, 2004.

Hewitt, Martin. 'Culture or Society? Victorian Studies, 1951–64', in *The Victorians since 1901: Histories, Revisions and Representations*, edited by Miles Taylor and Michael Wolff, 90–104. Manchester: Manchester University Press, 2004.

Hilton, Boyd. *A Mad, Bad and Dangerous People? England 1783–1846*. Oxford: Clarendon Press, 2006.

Hughes, T. Harold and E.A.G. Lamborn. *Towns and Town-Planning: Ancient and Modern*. Oxford: Clarendon Press, 1923.

Hunter, Michael, ed. *Preserving the Past: The Rise of Heritage in Modern Britain*. Stroud: Alan Sutton, 1996.

Ison, Walter. *The Georgian Buildings of Bath*. London: Faber and Faber, 1948.

Ison, Walter. *The Georgian Buildings of Bristol*. London: Faber and Faber, 1952.

Jenkins, Jennifer, and Patrick James. *From Acorn to Oak Tree: The Growth of the National Trust 1895–1994*. London: Macmillan, 1994.

Larkham, Peter J. *Conservation and the City*. London and New York: Routledge, 1996.

Mandler, Peter. *The Fall and Rise of the Stately Home*. New Haven and London: Yale University Press, 1997.

Matless, David. *Landscape and Englishness*. London: Reaktion, 1998.

McKellar, Elizabeth. 'Populism versus professionalism: John Summerson and the twentieth-century creation of the "Georgian"'. In *Articulating British Classicism: New Approaches to Eighteenth-Century Architecture*, edited by Barbara Arciszewska and Elizabeth McKellar, 35–56. Aldershot: Ashgate, 2004.

Meller, Helen. *Towns, Plans and Society in Modern Britain*. Cambridge: Cambridge University Press, 1997.

Newman, John. *The Buildings of Wales: Glamorgan*. London: Penguin Books, 1995.

O'Connell, Sean. *The Car and British Society: Class, Gender and Motoring 1896–1939*. Manchester: Manchester University Press, 1998.

Olsen, Donald J. *The Growth of Victorian London*. Harmondsworth: Penguin Books, 1979.

Parissien, Steven. *The Georgian Group Book of the Georgian House*. London: Aurum Press, 1995.

Parry, Graham. *The Trophies of Time: English Antiquarians of the Seventeenth Century*. Oxford: Oxford University Press, 1995.

Pendlebury, John. 'Planning the historic city: Reconstruction plans in the United Kingdom in the 1940s'. *Town Planning Review*, 74, 4 (2003): 371–93.

Pendlebury, John. 'Reconciling history with modernity: 1940s plans for Durham and Warwick'. *Environment and Planning B: Planning and Design*, 31 (2004): 331–48.

Rothnie, Niall. *The Baedeker Blitz: Hitler's Attack on Britain's Historic Cities*. Shepperton: Ian Allan, 1992.

Saint, Andrew. 'How listing happened', in *Preserving the Past: The Rise of Heritage in Modern Britain* edited by M. Hunter, 115–33. Stroud: Alan Sutton, 1996.

Searle, G.R. *A New England? Peace and War 1886–1918*. Oxford: Clarendon Press, 2004.

Sharp, Thomas. *Cathedral City: A Plan for Durham*. London: Architectural Press, 1945.

Stamp, Gavin. 'Origins of the Group'. *Architectural Journal* 175, 13 (1982): 35–8.

Strong, Roy. *Country Life 1897–1979: The English Arcadia*. London: Country Life Books and Boxtree, 1996.

Summerson, John. *Georgian London*. London: Pleiades Books, 1945.

Summerson, John. 'John Wood and the English town-planning tradition', in *Heavenly Mansions and Other Essays on Architecture*, 87–110. London: Cresset Press, 1949.

Sweet, Rosemary. *Antiquaries: The Discovery of the Past in Eighteenth-Century Britain*. London and New York: Hambledon and London, 2004.

Tarlow, Sarah. *The Archaeology of Improvement in Britain, 1750–1850*. Cambridge: Cambridge University Press, 2007.

Thrift, Nigel. 'The geography of late twentieth-century class formation'. In *Class and Space: the Making of Urban Society*, edited by Nigel Thrift and Peter Williams, 207–53. London: Croom Helm, 1987.

Train, Christopher. 'The growth of Victorian Ludlow'. In *Victorian Ludlow*, edited by David Lloyd, Roy Payne, Christopher Train, and Derek Williams, 32–52. Bucknell: Scenesetters, 2004.

Trinder, Barrie. *Beyond the Bridges: The Suburbs of Shrewsbury 1760–1960*. Chichester: Phillimore, 2006.

Urry, John. *The Tourist Gaze: Leisure and Travel in Contemporary Societies*. London: Sage, 1990.

Watkin, David. *The Rise of Architectural History*. London: Architectural Press, 1983.

Young, B.W. *The Victorian Eighteenth Century: An Intellectual History*. Oxford: Oxford University Press, 2007.

Chapter 9
Social Housing as Heritage: The Case of Byker, Newcastle upon Tyne

John Pendlebury, Tim Townshend and Rose Gilroy

Over the last decade or so in England, validation as cultural heritage has been extended to include the 'listing' of large groups of welfare-state housing. Selected principally on the basis of art-historical criteria, but significant also for their historical role as part of the mid-twentieth century approach to solving the housing problems of the working class, the validation of such estates presents new challenges and opportunities for the heritage sector.

This chapter is concerned with the potential impact of listing upon one such housing estate; Byker, Newcastle upon Tyne, parts of which are only about twenty years old. Famous nationally and internationally, many within Newcastle and Byker itself have a less positive view of the estate. Listing, eventually confirmed in January 2007 by the Department of Culture, Media and Sport, was initially proposed by English Heritage in 2000. In the meantime an active process of preparing for listing was undertaken with a variety of stakeholders including the principal landlord (which changed from being Newcastle City Council to the arms-length agency, *Your Homes Newcastle* in 2004), residents' groups and English Heritage. One major output of this was a 'conservation plan'. In parallel there have been efforts to address the significant social and physical problems that exist within the estate.

The focus in this chapter is the ways that Byker is valued both by those within the estate and by some of the professionals engaged in the listing of Byker as either housing managers or heritage professionals. In particular if people do consider Byker as somewhere unique and special, what does this mean? Does the recognition of the estate as special through statutory listing in anyway capture how the estate is valued?

The chapter is based upon interviews with a variety of stakeholders,[1] combined with documentary research. Interviews were undertaken primarily in 2005 and 2006, with some follow up interviews after the announcement of listing in 2007. Interviews took a variety of forms and some of the professionals were interviewed on a number of occasions. Resident interviewees were active in one or other of the community or residents' associations within the estate. Further interviews were

---

1   The authors wish to express their thanks to Trevor Wren who undertook some of the interviews on their behalf.

undertaken with community workers, local politicians, officers of *Your Homes Newcastle*, an architect from the original design team currently employed on an advisory basis by *Your Homes Newcastle* and heritage sector professionals (from English Heritage, Newcastle City Council and North of England Civic Trust).

The Heritage of Modernism and the Politics of Heritage

In 1979 'The Thirties Society' was formed reflecting a developing awareness and understanding of twentieth century buildings in the UK. Conservationists are frequently proud to see themselves as an avant-garde of changing tastes; the creation of the Georgian Group, Victorian Society and the Thirties Society (now the Twentieth Century Society) have all been considered as examples of the conservation movement anticipating, if not actively leading, shifts in public taste whereby architectural periods that were ignored or despised are positively re-evaluated.[2] In this context it is no surprise that the post-war architectural modern movement was reappraised and campaigning started for state protection of key buildings through listing,[3] notwithstanding the associations still fresh to many people of such buildings being responsible for the desecration of traditional townscapes in the preceding decades. Indeed, the case for conserving post-war public housing was the specific focus of a report by the Twentieth Century Society.[4] The strength of the conservation movement can be seen in its success in achieving the listing of buildings by the then Conservative Government that, in the late 1980s, were generally held to be deeply unpopular. More remarkable still, given the political hue of the Government, was the listing of council housing in the form of Alexandra Road Estate, Camden and Keeling House, Bethnal Green, both London, in 1993, at the time under threat of insensitive concrete repairs and a dangerous structure notice respectively. Admittedly little more public housing was listed until 1997 and a change in Government. Subsequently, a number of large-housing schemes have been listed in England, mostly in London. It is also noticeable that listing has focused on the architecturally distinct and thus far has not embraced the more conventional system built estates that are perhaps more representative of the post-war housing story. The conservation movement has therefore adjusted its parameters once again

---

2   See e.g. Stamp, G. 'The art of keeping one jump ahead: Conservation societies in the twentieth century', in *Preserving the Past: The Rise of Heritage in Modern Britain* edited by M. Hunter, (Stroud, Gloucestershire: Alan Sutton, 1996), 77–98; Hobson, E. *Conservation and Planning: Changing Values in Policy and Practice*, (Spon Press: London, 2004).

3   While, A. 'The state and the controversial demands of cultural built heritage: modernism, dirty concrete, and postwar listing in England', *Environment and Planning B: Planning and Design*, 34, 4 (2007): 645–663.

4   Bayley, R. *Celebrating Special Buildings: The Case for Conserving Post-War Public Housing*, (London: Twentieth Century Society, 2002).

to include a new category of buildings. However, this reading of such buildings as constituting heritage is not necessarily evident amongst a wider public. English Heritage has commissioned a series of opinion polls from MORI[5] which have asked people what images the term heritage conveys. The most popular responses focus around fairly 'traditional' types of heritage such as historic buildings, stately homes and historic parks and gardens. Few respondents considered modern buildings to be heritage, even when prompted.

This demonstrates that ideas of what constitutes heritage are neither universal nor fixed. That heritage is socially constructed has been a truism in much academic debate and has been increasingly acknowledged in practice arenas. For example, work by the Getty Institute has asserted the idea that the notion of heritage is universal, but is articulated in culturally specific ways.[6] In the UK, English Heritage has acknowledged the multiplicity and fluidity of heritage values that exist, some of which relate to culturally-specific conditioning.[7] Academic literature has often taken a more critical approach to these issues. In the same way that people's perceptions of heritage tend to focus on the constructions of the elite, it has been argued that the mobilization of the concept of heritage is a political tool of certain classes in society, essentially a culturally-focused part of the upper and middle-classes. Critiques range from grand theory, such as dominant ideology theory,[8] to explorations of more practical politics, such as using cultural heritage to prevent development, the so-called NIMBY[9] phenomenon. A recent contribution to this debate has been made by Laurajane Smith, with her concept of the 'Authorized Heritage Discourse'.[10] Central to this is the argument that ideas about heritage are developed and controlled by elite actors, for example through conservation charters or formal mechanisms of protection such as listing, in ways which are said to override 'subaltern' concepts of heritage, including those which might arise from the community.

The heritage sector has become acutely conscious of the need to escape its elitist image, especially since the election of a modernizing Labour Government in 1997 that has shown no great interest in heritage issues.[11] In this context the

5   MORI, *Attitudes Towards the Heritage*, (London: English Heritage, 2000); MORI, *Making Heritage Count?* (London: English Heritage, DCMS & Heritage Lottery Fund, 2003).

6   Avrami, E., R. Mason, et al., *Values and Heritage Conservation*, (Los Angeles, The Getty Conservation Institute, 2000)

7   English Heritage, *Conservation Principles: Policies and Guidance for the Sustainable Management of the Historic Environment*, (London, English Heritage, 2008).

8   See, for example, Merriman, N., *Beyond the Glass Case*, (Leicester: Leicester University Press, 1991).

9   Standing for 'Not In My Back Yard'.

10   Smith, L., *Uses of Heritage*, (Routledge: London, 2006). See also Smith chapter this volume.

11   Pendlebury, J., Townshend, T. and Gilroy, R. 'The conservation of English cultural built heritage: A force for social inclusion?', *International Journal for Heritage Studies*, 10, 1, (2004): 11–32.

expansion of post-war listing, including large estates of social housing, assumes some significance. Is this valorization part of wider democratic and inclusive ideas of how society might collectively think about heritage, or is it purely a top-down construction of a cultural elite, part of an 'Authorized Heritage Discourse' imposed on residents? We return to this question in the context of Byker at the end of the chapter.

The Re-development of Byker

In 1968 Ralph Erskine was appointed architect and planner for Byker, a remarkably radical appointment by the Conservative controlled local authority. Erskine had the view that the physical fabric of place could be transformed without disrupting relational resources. This was a step change from clearance elsewhere, where the focus was on housing as defective bricks and mortar rather than housing as home and a place of attachment. Thus, fundamental to the redevelopment of the streets of Tyneside flats that made up Byker was the strong wish to preserve the social character of the area. What that character and spirit was probably differed little from that of working class communities across the region where families lived close to the industries that paid their wages; where girls married the boys they went to school with and set up home in the same street as their parents. What was different was that this character was recognized and a sympathetic response made.

The bulldozers arrived in the late 1960s and moved around the neighbourhood for a decade. Work started at the north end of the area on a site intended for an urban motorway between the major shopping street, Shields Road, and the site. Erskine's solution was to create a perimeter block incorporating housing that would both act as a sound barrier and make a strong visual signal to all that behind the wall was a separate place – a village in the city – with few pedestrian and vehicular access points through. This is the 'Byker Wall' (Figures 9.1 and 9.2). Behind this was constructed a new street layout with south facing communal courtyards replacing the old grid iron pattern. These spaces were filled with planting boxes, seats and tables where neighbours could sit and talk (Figure 9.3). The dwellings themselves were low rise of varied size to fit the needs of a whole community and characterized by light materials – timber cladding often brightly coloured and metal roofs; architecturally the style has been called 'romantic functionalism'[12] or 'romantic pragmatism'.[13] The integration of landscaping into the estate was a key integral feature, including so called 'ruin

---

12   North East Civic Trust, *A Byker Future: The Conservation Plan for the Byker Redevelopment, Newcastle upon Tyne*, (Newcastle upon Tyne: English Heritage & Newcastle City Council, 2003).

13   Department Of Culture Media & Sport, *Planning (Listed Buildings And Conservation Areas) Act* 1990, 71st Amendment Of The 7th List Of Buildings Of Special Architectural Or Historic Interest, City Of Newcastle Upon Tyne (Tyne & Wear), (London: DCMS, 2007).

Figure 9.1   The 'outside' of the Byker Wall from Dalton Street
Source: © Rose Gilroy

Figure 9.2   The southern 'inside' of the Byker Wall from Raby Way
Source: © Rose Gilroy

Figure 9.3   A typical courtyard space
Source: © Rose Gilroy

**Figure 9.4   The significance of landscape and retained landmark buildings
          – view to St Michael's church**
Source: © Rose Gilroy

Figure 9.5   'Ruin bits'
Source: © Rose Gilroy

bits'; architectural fragments from demolished buildings across the city (Figures 9.4 and 9.5). Cars were mostly kept at the periphery of the estate and a number of older buildings were retained, such as, churches and pubs. New social facilities introduced included a large number of 'hobby rooms' for clubs and individuals to use. Erskine's practice set up office in a former undertaker's premises in the heart of the area where their open door policy allowed a demystifying of the architectural process.[14] However, this was not community architecture in the sense of residents having a strong role in design – indeed it has been noted by Abrams how strongly the design relates to other work by Erskine.[15]

The rolling programme of clearance and rebuilding attempted to reduce the displacement of residents who would see their new homes being built, then move into them and see their old homes demolished. The City Council as landlord also worked to make its practices fit with the needs of the Byker community – so

---

14   See Drage, M. 'Byker: Surprising the colleagues for 35 years, a social history of Ralph Erskine's arkitektkontor AB in Newcastle', *Twentieth Century Architecture: The Journal of the Twentieth Century Society*, 9, (2008): 147–162, for a recent account of the construction of Byker from one of Erskine's architectural team.
15   Abrams, R. 'Byker Revisited', *Built Environment*, 29, 2, (2003): 117–131.

dwellings were pre-allocated with a strong concern for facilitating the retention of neighbourly contacts that were seen to be supportive. The success of this policy can be over emphasized; ultimately less than half of the old Byker residents returned. The redevelopment was also dogged with problems – industrial action in the 1970s caused delays and the introduction of the Right To Buy (*Housing Act* 1980) coupled with a large scale moratorium on local authority house-building meant that by the early 1980s two large sites were still vacant with no possibility of being developed according to the original plan. The Byker redevelopment ultimately comprised 2,010 dwellings and today houses around 9,500 residents.

Byker Today

The southern part of the Byker redevelopment suffered problems of vandalism from the early 1980s. The greater problems of south Byker have been the subject of much speculation; for example, as this part of the estate rehoused far fewer Byker residents did new tenants see it simply as another council estate and not a place to call home? Decline in this part of the estate reached a head with proposals to demolish a small but architecturally significant area called Bolam Coyne in the late 1990s (Figure 9.6). Such a drastic intervention in a housing estate that has won many national and international plaudits immediately triggered controversy within the estate, the city and nationally.

Subsequently, in 2000, English Heritage recommended that the Government list the entire Erskine development at grade II*. The Department of Culture, Media and Sport took nearly seven years to reach a decision, although in 2003 it did announce its intention to list as part of a public consultation (albeit on the basis of a lower status grade II listing).[16] In the meantime planning went ahead on the basis that listing would happen. Thus a Conservation Plan was commissioned for the estate, undertaken by the North East Civic Trust with extensive community participation (Figure 9.7).[17] In 2004 two areas in South Byker were subject to a design competition, both incorporating vacant sites and existing housing stock. One of these directly incorporated most of the Erskine estate south of Commercial Road, including Bolam Coyne. The result of the competition was announced in 2006, since when there has been a further hiatus of action. People involved with Byker have also looked outwards; for example, there have been reciprocal visits with the listed Park Hill Estate in Sheffield, a pioneering deck-access housing scheme of a very different architectural character. Finally, the entirety of the Erskine designed-estate was listed at grade II* in January 2007 (the listing excludes pre-existing buildings incorporated into the design with the exception of the building that housed Erskine's office).

---

16   Buildings can be listed as grades I (the highest grade), II* or II. The higher grades are used sparingly with over 90 per cent of buildings being grade II.

17   North East Civic Trust, *A Byker Future* (see note 12)

Figure 9.6   The northern outside edge of Bolam Coyne in 2008
Source: © Rose Gilroy

Figure 9.7   The cover of the Byker Conservation Plan, *A Byker Future*
Source: © North of England Civic Trust

Thus after a long hiatus DCMS has validated the Erskine designed estate as 'particularly important buildings of more than special interest'.[18] However, other recent appraisals of Byker have been less positive. For example, two non-local academic commentators with long Byker associations have both written of it as a 'failed' project. Ravetz, who lauded the estate in 1975, wrote 'the potential listing of Byker is, at the time of writing, finely balanced against its demolition as a failed estate',[19] over-dramatizing the real problems that exist. Similarly Abrams reappraisal, on the basis of periodic visits from the USA, referred to shock at the decline of the estate, suggested conspiratorial housing management policies to provoke crisis (to then secure more resources) and concluded with an environmentally deterministic evaluation of the design concepts used; '(the) threat lies in the rigidity of the architect's preconceived vision for the community. It was sheltering, too centrifugal... it unintentionally had a medieval degree of separation from the surrounding community... effective in preventing a sustainable community from developing'.[20]

The responses of our interviewees talking about Byker today picked up some of the negative features of the estate suggested by Ravetz and Abrams, but also many of the positive facets of Byker, as might be associated with somewhere enjoying listed status. Any characterization of the views of a community of nearly 10,000 residents will inevitably be a simplified account. However, there were conceptions of Byker recurrent through the different categories of respondents we talked to which are important in understanding how residents and professionals perceive it.

A dominant theme throughout all our interviews was the sense of not one but several Bykers and how this is often articulated through the spatial differentiation of different parts of the estate. Broadly speaking people would refer to the Byker Wall, 'middle Byker' (south of the Wall, north of Commercial Road – the only through road bisecting the major part of the estate) and 'south Byker' (south of Commercial Road). Other areas were sometimes significant in discussion, for example Dunn Terrace, physically separated to the west. These geographical distinctions were not neutral – they each were used to indicate different sorts of communities. 'Middle Byker' was regarded as the heart of 'old Byker' in the sense that this was where there is a concentration of long-standing Byker families and a sense of the Byker community. The Byker Wall, whilst having some particular problem flats on the lower floors and accessed directly off stairwells, was largely considered a successful community but of a rather different nature with incipient colonization and gentrification by a young bohemian group including professionals and artists. South Byker was consistently regarded as the most problematic part

18   Department for Communities and Local Government, *Revisions to Principles of Selection for Listed Buildings*, (TSO, Circular 01/2007, 2007), Para 6.6.

19   Ravetz, A., *Council Housing and Culture: The History of a Social Experiment*, (Routledge: London, 2001), 182. See also comments by Malpass, this volume.

20   Abrams, R. 'Byker Revisited', p. 130 (see note 15).

Table 9.1   Interviewees referred to in this chapter

Conservationists	H1	North of England Civic Trust officer
	H2	Conservation Officer with Newcastle City Council
	H3	English Heritage officer
	H4	English Heritage officer
	H5	Conservation Officer with Newcastle City Council
Housing	Y1	*Your Homes Newcastle* Housing Officer
Management	Y2	*Your Homes Newcastle* Housing Officer
	Y3	*Your Homes Newcastle* Housing Officer
	Y4	Architect from original design team advising Your Homes
Community workers	C1	Community worker, living in Byker
Residents	R1	Lifelong Byker resident from Dunn Terrace area
	R2	Resident in Wall
	R3	Resident
	R4	Retired lifelong Byker resident
	R5	Young professional resident in Wall
	YMCA group	Group of young Byker residents

of the estate, characterized by a lack of long-standing residents, by a high tenancy turn over, high rates of empty property and the increasing use of the area to house asylum seekers.

The belief that Byker has an unusually strong sense of a surviving place-bounded community was a powerful feature of our interviews with both residents and those professionals who had significant community contact in Byker. It came out as a distinctive feature of the area almost as much as the design. There was some speculation about why this might be the case; one element of which was thought to be related to a strong tradition of neighbourhood organization and representative groups. However, the sense of some very real problems for residents on the estate came through all our interviews and the Conservation Plan reported the ward as being the then 78th most deprived ward in the country. Mostly these were focused around drugs and anti-social behaviour, the latter principally from groups of young people but also through changing attitudes towards neighbourliness and the public realm. Again, distinctions were made between different parts of the estate, with the perception that social problems were worst in the southern part of Byker.

A number of our interviewees remarked that peoples' experience depended upon how much choice they had had in living there. Thus old Byker residents or some of the new professionals were much more likely to have a positive feeling about the estate than those housed with little choice by the Council/*Your Homes Newcastle*. There was also a feeling that Byker has a poor image in the rest of Newcastle. For example,

> Byker's got a really bad reputation from outsiders... but if you live here its
> totally different, it's a great place to live, it's really community orientated, it's no
> worse than anywhere else. The houses are lovely, most people are really friendly,
> great community spirit, everybody mucks in together... (interviewee R1).

Despite the evident problems that have befallen Byker a number of respondents
felt, at the time they were interviewed, that the estate was getting better,

> Five to six years ago Byker was probably on its way to being a sink estate,
> turnovers, length of leases were very short, turnovers high, high voids rates,
> areas where there was a level of social problems and those areas had been lost in
> many ways. If you looked at them you'd say the areas been given up on... [Now]
> Byker is almost at full capacity and we've turned around the high turnover rate,
> credit partly to the housing officer.. (interviewee R5).

It is invidious to separate discussion of community cohesion and physical
environment. People's description of the two was often strongly interwoven. This
was reflected in both positive and negative appraisals of design and community.
On the positive side, 'My development is on a square and all the bairns play in
the square, totally safe and everybody's looking at the windows and watching as
they're playing so they're safe as houses' (interviewee R4). However, alternatively,
'the way Byker is designed is very vulnerable to anti-social behaviour, you just
need one junkie in one house and that whole terrace is quickly going to be under
threat of deterioration' (interviewee R5). And a respondent who saw the strength
of the design working both ways,

> There's the wall which defines the area, the actual Byker Wall itself which actually
> helps to stigmatize the place I think in that folks from outside don't want to step
> beyond that barrier and those that do pluck up the courage to venture in are
> surprised by what goes on in here because, one thing that I'd like to say, is that its
> got a village feel to it, its like an urban village and that again is reinforced by the
> wall so because the boundaries are so clearly defined... That sense of village and
> identity of the place and there is a sense of community here which you wouldn't
> necessarily find in other places. There are a lot of problems associated with the
> area from the point of view of social deprivation and crime so the reputation the
> place has got is deserved in some ways but exaggerated and reinforced by the
> design of the place. I find it an interesting place to live, from the point of view
> that there's quite a lot of green space, the place has been planted with quite a lot of
> shrubbery and trees which aren't found on other council estates in the area. Also
> the way in which its been designed makes it interesting from the point of view that
> nowhere looks the same. You can stand in any particular place and it will always
> look different because it's not your standard of row upon row of houses, the design
> adds to the interest of the place. It also adds to the problems of the place in that it's
> difficult to police and there have been lots of problems associated with that. There

are lots of alleyways, mainly pedestrianized so it's not easy for vehicular policing and it's very easy to hide away very quickly (interviewee C1).

The role of landscape on the estate was an issue that divided our resident respondents between those who saw it as a great strength of Byker and others whose key association was landscape as a cover for crime. For example,

> The architects listened to what the residents had to say and incorporated into the designs the ideas the residents put forward so you had a tremendous amount of trees and shrubs, which we never had. The bird life in Byker now is phenomenal (interviewee R4).

Alternatively,

> Oh definitely, they [the tenant's association] wanted more bushes and trees whereas we don't, we think there's far too many. At Rabygate... through the archway there's three bits of grass and at one time that was all full of bushes and there's paths going up and old people that's next to the bushes were getting mugged so we fought them, we got them taken out. (interviewee R2).

A consistent view advanced by the heritage professionals was that, despite the problems Byker has endured, they had been pleasantly surprised by the overall condition of the estate, viewing physical problems of fabric as being neither profound nor systemic. This appraisal is supported by the work undertaken for the Conservation Plan. This embraced both basic building condition and the survival of the original design philosophy. Indeed, the heritage professionals were very positive about the way those officers from the City Council, and subsequently *Your Homes Newcastle*, had recognized the unique qualities of Byker and sought to sustain them in difficult circumstances as resources were lost from an estate with a high maintenance requirement. In terms of physical problems, it was felt that the landscape has suffered much more, the reasons for which are bound up in a combination of issues to do with vandalism and maintenance.

However, there was an acknowledgement from the heritage professionals that the problems of the area were connected with its design. A strong theme was a feeling that the design of the spaces in Byker might have been suitable in the 1970s but the changing nature of society had made the design much more problematic. This overall trend is what the Conservation Plan identified as a 'collective retreat'. In essence this describes a process whereby residents' focus is now much more on their own private domain of house and garden, with the consequent semi-abandonment of many of the informal semi-public, semi-private spaces that are integral to the design of Byker. This focus on the individual domain, rather than the collective, is also evident with practical issues, such as rising levels of car ownership. This has caused problems in a largely pedestrianized estate, with much authorized car-parking confined to peripheral car parks with poor security. Not surprisingly in these circumstances car owners have found ways to get their cars nearer to their property.

In summary, despite some significant problems, most of our respondents from all groups had an essentially positive view of Byker today, both in terms of the strong sense of community felt to exist through much of the estate and in terms of the physical environment. Indeed, universally our interviewees viewed Byker as somewhere 'different' and 'special'. The next section further considers 'insider' (i.e. those living and working on the estate) and 'outsider' and official (i.e. non-resident professionals) views of what makes Byker special.

Byker is Special

The official articulation of this is set out in the Government's 'list description'. This document, issued in January 2007, runs to over 100 pages; which is not surprising as it effectively contains a summary architectural description of the whole estate. Nevertheless, whilst there is a brief overall statement of significance and some background history, it is principally an extended version of an 'old-style' listing document,[21] focusing on architectural description and containing relatively little interpretative material on what it is about Byker that warrants grade II* listing.

The other formal, but more discursive, document that considers the qualities of Byker is the Conservation Plan.[22] Following standard Conservation Plan methodology[23] the plan includes a clear statement of significance for the estate. The preamble makes it clear that the process of development is as important as the material product. Significance is considered under the following briefly summarized headings (the full analysis runs to over fifty pages):

- Architectural: 'as a thorough, holistically-designed composition in a confident and informed "romantic functionalist style"'
- Landscape: 'was thoroughly integrated with the overall masterplanning of the Estate, an unusual approach for the time, creating an enduring link between the housing, the people and their environment'
- Social and community: 'social and community significance is ingrained in the Estate's fabric and in the innovative processes which helped create it. Byker's history, its people and their hands-on involvement in designing the

---

21   As part of the review of the heritage protection regime English Heritage has shifted to preparing fuller and more interpretative list descriptions, with clear statements of significance. Byker follows the old style of description, presumably as it was prepared in 2000, long before the new style was developed.

22   North East Civic Trust. *A Byker Future* (see note 12). See pages xxi and xxii for quotations.

23   Kerr, J.S., *The Conservation Plan: A Guide to the Preparation of Conservation Plans for Places of European Cultural Significance*, (New South Wales National Trust, 1996). Clark, C., *Informed Conservation: Understanding Historic Buildings and their Landscapes for Conservation*, (London: English Heritage, 2001).

place are crucial to understanding its importance... Byker's community is social, active and vocal'

- Historical: 'it was at the forefront of a sea-change in UK social housing towards more humane architecture than its often brutalist predecessors. Its direct influence can be seen in much housing since, including current schemes'

- Townscape: 'It responds firmly to the topography and politely to neighbours, and its dramatic and intriguing presence give it an unmistakeable identity on a wide scale...'

- Archaeological and ecological significance are also mentioned in passing.[24]

Many of the heritage professionals made similar analyses though in more personal terms. For example,

> I think of it more on the smaller scale stuff than the bigger scale stuff... I think it's very romantic, what intrigues me more is the more you go round it, the more romantic and picturesque it seems, almost like an eighteenth century landscape, you approach and suddenly you see the alleyways are leading up to a focal point on the wall or a church or something like that. The way that works I think is superb, but that's not to ignore the way that some parts of the estate is pretty awful (interviewee H4).

The way insider interviewees talked about what makes Byker different and special was rich and complex. They talked about qualities such as issues of time and attachment to place, community, kinship and propinquity and the qualities of place in terms of the Erskine design and the housing stock. The North East Civic Trust in their consultation work for the Conservation Plan asked people what is special about Byker. Replies included 78 per cent of respondents who felt Byker to be special because 'it is not like typical terraced housing elsewhere', 72 per cent because 'there are lots of shrubs and trees' and 67 per cent because 'of the variety of building materials and colours'.[25] Other ways the particularity of the estate was mobilized included the evident appreciation of Byker by others, as demonstrated by the visitors the estate still receives from across the world, and by external reference such as the reciprocal visits to the unfavourably compared Park Hill.

Though they often use different language, ultimately both residents' and professionals' descriptions of Byker interweave issues of community and of design. The way residents talk about Byker tends to foreground issues of community in terms of its strengths and its problems. But the importance of the

---

24   Quotations from North East Civic Trust. *A Byker Future* (see note 12), 144–145.

25   The Byker Conservation Plan was underpinned by a highly unusual degree of community engagement with many meetings, a visit to Park Hill in Sheffield etc. These particular figures are based upon c. 80 questionnaire responses.

nature of the environment as a physical frame for social interaction was a major theme. Positively this might mean, for example, the way the design facilitated neighbourliness and safe children's play and a delight in the rich landscape compared to 'old Byker'. More negatively it might mean vandalized spaces cared for by no-one and landscape cover for muggers. The heritage professionals not surprisingly focused more on the qualities of physical design. However, again issues such as the hierarchy of open space and the importance of the landscape were very prominent in their descriptions. Furthermore, the role of community both in terms of the formation of the estate and contemporary management was prominent in discussion. This is most conspicuously absent in the formal listing description.

Perceptions of Listing

Most of our interviews and all those with residents were undertaken whilst a decision on the listing was still pending. The proposed listing of Byker had been controversial within the estate. Despite the pride many residents have in Byker this does not mean the proposed listing was met with acclaim. Indeed, though residents saw the estate as special, listing was viewed with a great deal of suspicion. In terms of the people we interviewed there was no clear correlation between attitudes to listing and their general feelings about Byker; it was noticeable that two of the interviewees most effusive about the qualities of Byker had the least support for the idea of listing. Our heritage sector interviewees, whilst acknowledging the reservations in the area over listing, felt that this was largely based on misunderstandings of the implications of listing, often fuelled by rumours. Listing was, they felt, equated with a strict preservation approach and therefore perceived to be an impediment to improving the estate. Part of the rationale behind the commissioning of the Conservation Plan was to overcome these fears. For example,

> Local people were trying to find ways to celebrate Byker but they couldn't make the connection in their heads between the opportunities that listing provides, all they could see, all they had heard about and all the grapevine was talking about were the potential negative aspects of it (interviewee H1).

In similar vein,

> One of the first misconceptions about the listing was when the new management review came in for lockable new doors on the wall, new security doors, that somehow that wouldn't be allowed because they weren't Erskine designed, and all the tenants in the wall thought EH were stopping perfectly reasonable secure access and so on and it was that kind of thing that we were being misrepresented over (interviewee H4).

In the interviews with residents we undertook, most of whom were fairly well connected into various networks in Byker, it was clear that few had much understanding of the listing process. One of the widest spread misconceptions, circulated also by the local media, was that the estate had indeed been listed. Equation of listing with strict preservation was also evident. For example,

> If you list it, it can't be ever moved. I think it shouldn't be listed. What if they want to do new ideas, and want changes and people want to make new houses and that? If you've got your own property you can't do anything you want, you've got to stay in [with what] the Government conforms and that (YMCA group).

However, residents and others did also talk about positive benefits listing might bring, both materially and otherwise. Much of the debate in early discussions on listing had been over the grading that might be attributed to Byker. English Heritage had suggested grade II* but DCMS consulted on the basis of grade II. A number of our interviewees expressed a view that a momentum had developed, especially within the Council, that if listing was to occur it should be at grade II*. In part this was attributed to issues of local pride but more materially it was based upon the perception that the higher grade might be a lever for securing resources from such as the Heritage Lottery Fund.

Beyond the direct capture of resources listing was seen by some as a potential means of developing the interest in Byker from those outside the estate. For example,

> I wonder if it was listed and properly promoted whether there wouldn't be a small amount of tourist trade to be gained? Could mean employment for half a dozen people here but also lift people's views of where they're living; it all depends on how its sold, with a positive spin, this is something really good for Byker that's going to put Byker on the map, make where you live a good place to be... (interviewee C1).

The potential image benefits were mentioned by another interviewee,

> It could be a positive, yes. You could get city officials sticking their chests out and saying we're proud of the fact we've got a heritage listed estate in our city, you know. It's a plus point for those who are in command in the city; it would certainly improve people's outlook on, let's face it Byker is a working class area which has basically been looked down on... (interviewee R4).

The potential positive benefits of listing were implicit in much of what our heritage professional interviewees had to say. In terms of the housing managers we interviewed there was an interesting translation of the particular mechanism of listing to a wider discussion of specialness. For example,

Listing is a bizarre concept but if you ask people about what is good and bad about the estate they will tell you. So we approached the public consultation. Let's assume that Byker is going to be around for a long time what would you like to keep and what would you like to change? It's a completely different debate than would you like it to be listed. We felt it was not an intellectual debate but one hinged on Byker is a good place designed with the best of intentions, let us understand how it got there, the intentions, the problems, the risks etc. I was reading an article in *The Guardian* the other day by Peter Hetherington who was criticizing the design generated by Pathfinder and he was saying why not go back and look at something that works such as Byker – its about architecture and its relationship to community (interviewee Y1).

And from one of the original design team employed by *Your Homes Newcastle* to advise on recent works,

I think we can't preserve it in aspic and the way it was isn't necessarily going to work now in terms of car parking – again so treating it with some understanding and respect is what I think it should mean which is what the Conservation Plan was about- so that we could make sympathetic judgements about issues (interviewee Y4).

However, as well as the reservations about the particularities of listing, and the degree to which it might frustrate improvements to the estate, there was evidently some more profound distrust of the process reflecting a more generalized suspicion of authority. In the particular circumstances of Byker this translated to empty promises over improvements which might be made to the management of the estate. For example,

...a number of people were very angry at the meeting and felt very abandoned and turned up with pictures of the state of the housing, the vandalism, the record of problems they'd suffered and that's the management not achieving a level of management which they perceived quite rightly that they deserved and so they'd lost confidence (interviewee R5).

Specifically interviewees talked about Bolam Coyne. It was the threat to Bolam Coyne that triggered off the listing proposal initially and at the time of writing it remains fenced off and derelict. Whilst there may be a range of factors that have led to this, hardly surprisingly there was a tendency to relate this inertia to the listing proposal. For example,

... and it can't be demolished and people just can't understand why that is the case and I think in some sense Bolam Coyne might be a touchstone, people will think oh well if its going to be listed we have to keep Bolam Coyne, we can't for the life of us see why we have to keep that... (interviewee R3).

And,

> [Bolam Coyne] almost sits there as an open wound and reminds people all the
> time that this place is a failing place, so from that perspective I've always felt
> something needed to be done about it, either demolishing or upgrading but
> something positive done in that space for me would help to heal some of that
> wound and start to turn around peoples' opinion of the place in which they lived
> (interviewee C1).

The lack of action on Bolam Coyne, notably to demolish, is said to have led to the
disbanding of one residents' association.

The Value of Listing

Ultimately listing did not come across as an especially important issue to our
interviewees who were residents, other than insofar that it has the potential to
frustrate improvements to Byker or perhaps to bring some modest benefits. This
is likely to be true for the vast majority of people in the area, other than perhaps
for a handful of champions for the conservation cause. Out of our interviewees
even the one with most commitment to listing Byker, and to conservation issues
generally, remarked,

> ...to most people... I don't think the listing matters, I think its about sorting out
> the issues of managing Byker. I want to see Byker succeed but not to the expense
> of the people living in the communities... the only interest I think with regards
> to listing would be would this improve our quality of life, that ultimately comes
> back to can it be managed better? (interviewee R5).

So what will the listing of Byker mean in practice and how will it affect the
management and evolution of the estate? Will it be considered as an architectural
monument, with an emphasis upon it architectural value and fabric, or will the
listing provide a broad interpretive framework through which a discourse about
evolution and change might be mediated? How will core conservation concepts
such as authenticity be judged? Whose values will prevail?

The Conservation Plan was an important benchmark in these debates and
sought to straddle the various competing agendas that exist in Byker, but this was
only the start of an ongoing process. Simplifying, there seems to be two broad ways
in which the listing of Byker might come to be seen. First, listing might be seen
as an outside imposition which frustrates the aspirations of residents and causes
inertia and delay. Whilst all our heritage and housing professional interviewees
talked very differently about the nature of listing, with the conservationists being
particularly anxious to avoid being tainted with preservationism, it is hardly
surprising that listing is perceived in this light. The listing debate was started by

the proposal to demolish Bolam Coyne and many years later it still stands derelict and fenced off as a festering sore and representative, for some at least, of what listing might mean.

However, the more positive future for listing might be as a rhetorical device; as a means of changing the nature of debate over the physical management of Byker that makes all the multiplicity of people and agencies involved acknowledge that it has particular and different qualities. It *might* force the various agencies to think beyond standard approaches or knee-jerk responses to problems. Furthermore, it *might* be the stimulus for some modest creative activity, such as providing an impetus to develop resources for learning about the estate both for inhabitants and visitors and it *might* be helpful in addressing some of the physical issues on the estate, such as landscaping, that won't be addressed in the substantial investment taking place to improve housing conditions through the Government's *Decent Homes* programme.[26] If listing can be combined with a better environment and better management practices it might just have some positive role in reinforcing a better self-image for the estate. It might possibly be part of a process of creating a better image for Byker for those beyond the area.

One of the difficult issues that lies behind this, however, is the perceived 'ownership' of listing. Ultimately listing is a highly bureaucratic, top-down process and how it is conveyed and enacted will need very sensitive handling if it is to have local credibility. Unresolved issues such as Bolam Coyne are tremendously damaging in this regard. Conversely if there is positive encouragement for colonization of the area in whole or part by an outside population, as a management device for 'improving' the area and better looking after the heritage, this might equally be perceived by the current community as a threat. The conservation community in the UK has tended to be blind to the consequences of gentrification. Certainly attitudes towards modernist housing have changed in London and many of the post-war listed estates have been gentrified (interviewee H3). Whilst London might be a special case it is noticeable following earlier failures to secure heritage money to improve Park Hill the problems of the estate are now being addressed by its wholesale transformation/ gentrification by the developers *Urban Splash*.

Conclusion

Byker's physical design is an integral part of what makes it what it is. It is a very particular and different place which alternately causes delight, fear or an

---

26   This is the new minimum standard of housing conditions for all those who are housed in the public sector. The Government set out a target in 2000 that it would ensure that all social housing meets set standards of decency by 2010, and public housing bodies have been required to set out a timetable under which they will assess, modify and, where necessary, replace their housing stock according to the conditions laid out in the standard. Criteria focus on repair, modern facilities, e.g. in kitchens and bathrooms, and thermal comfort.

image of a place apart, not to be ventured into; in this respect Abrams was right to focus on the significance of design if not in her conclusion that this inevitably leads to a failed community.[27] The importance of its physical characteristics was evident in all our interviews; whether people were praising or lambasting Byker its physical attributes were wound into their narratives. But Byker's architecture and landscape is not all of what it is or what makes it special. There is a consciousness of the history of a community and how that (in part) was involved in a process of redevelopment that led to continuity in space whilst the nature of the place was utterly transformed. This history is self-consciously recalled in the present. It is embedded in the institutions of neighbourhood governance and civil life and in the networks from that early period that still survive (in part) of kinship and friendship.

All this contributes to a sense that Byker is special. Clearly this would not be a universal view and there will undoubtedly be residents who passionately hate the place, but there does seem enough evidence to suggest that this identification of specialness is credible. Listing does not articulate what residents feel is special about Byker, because for all the recognition by heritage professionals of the importance of the community listing is defined by 'architectural or historic interest'. However, this does not mean discourse about the history of Byker is entirely controlled by an externally imposed 'Authorized Heritage Discourse', nor that listing has no local relevance. Whilst 'expert' views of the significance of Byker certainly have a significant impact on how Byker is talked about and managed this forms part of a more complex narrative. Not only do Byker people seem to think of the place as different and special but the way heritage professionals think about Byker, at local level at least, is mediated by their engagement with the various communities within Byker. And whilst listing does not necessarily align with the values that underpin these broader conceptions of specialness, maybe it is the best proxy measure we have. Ultimately though it is not so much the act of listing that it is important, it is what happens now. The creation of Byker was lauded for its process as much as its architecture. The success or otherwise of its listing will be the same.

Bibliography

Abrams, R. 'Byker revisited'. *Built Environment,* 29, 2, (2003): 117–131.
Avrami, E., R. Mason, et al. *Values and Heritage Conservation.* Los Angeles, The Getty Conservation Institute, 2000.
Bayley, R. *Celebrating Special Buildings: The Case for Conserving Post-war Public Housing.* London: Twentieth Century Society, 2002.
Clark, C. *Informed Conservation: Understanding Historic Buildings and their Landscapes for Conservation.* London: English Heritage, 2001.

---

27   Abrams, R., 'Byker Revisited', 130, (see note 15).

Department for Communities and Local Government. *Revisions to Principles of Selection for Listed Buildings*. TSO, Circular 01/2007, 2007.

Department of Culture Media & Sport. *Planning (Listed Buildings And Conservation Areas) Act 1990, 71ˢᵗ Amendment Of The 7ᵗʰ List Of Buildings Of Special Architectural Or Historic Interest, City Of Newcastle Upon Tyne (Tyne & Wear)*. London: DCMS, 2007.

Drage, M. 'Byker: Surprising the colleagues for 35 years, a social history of Ralph Erskine's arkitektkontor AB in Newcastle'. *Twentieth Century Architecture: The Journal of the Twentieth Century Society*, 9, (2008): 147–162.

English Heritage. *Conservation Principles: Policies and Guidance for the Sustainable Management of the Historic Environment*. London: English Heritage, 2008.

Hobson, E. *Conservation and Planning: Changing Values in Policy and Practice*. Spon Press: London, 2004.

Kerr, J. S. *The Conservation Plan: A Guide to the Preparation of Conservation Plans for Places of European Cultural Significance*. National Trust: New South Wales, 1996.

Merriman, N. *Beyond the Glass Case*. University Press: Leicester, 1991.

MORI. *Attitudes Towards the Heritage*. London: English Heritage, 2000.

MORI. *Making Heritage Count?* London: English Heritage, 2003.

North East Civic Trust. *A Byker Future: The Conservation Plan for the Byker Redevelopment, Newcastle upon Tyne*. Newcastle upon Tyne: English Heritage & Newcastle City Council, 2003.

Pendlebury, J., Townshend, T. and Gilroy, R. 'The conservation of English cultural built heritage: A force for social inclusion?', *International Journal for Heritage Studies*, 10, 1 (2004): 11–32.

Ravetz, A. *Council Housing and Culture: The History of a Social Experiment*. Routledge: London, 2001.

Smith, L. *Uses of Heritage*. Routledge: London, 2006.

Stamp, G. 'The art of keeping one jump ahead: conservation societies in the twentieth century'. In *Preserving the Past: The Rise of Heritage in Modern Britain* edited by M. Hunter, 77–98. Stroud, Gloucestershire: Alan Sutton, 1996.

While, A. 'The state and the controversial demands of cultural built heritage: modernism, dirty concrete, and postwar listing in England'. *Environment and Planning B: Planning and Design*, 34, 4 (2007): 645–663.

Chapter 10
Whose Housing Heritage?

Peter Malpass

This chapter explores some of the issues and tensions around housing and heritage. A house is primarily a place in which a succession of people will make their homes, and in doing so they will naturally want to be able to modify the physical structure in line with changing needs and preferences. However, this can be challenged by those who argue that the architect or designer has some ownership of the design and therefore some right to the protection of its integrity. There is also the question of responsibility to future generations: the durability of houses makes them rather like land, and so there is an argument that each generation has a responsibility to hand on the housing stock in at least as good a condition as it found it. This implies not just maintenance but also improvement, modernization and possible sub-division and extension of existing dwellings.

While there are some aspects of debate around housing and heritage that apply across tenures, attention in this chapter is directed towards a particular subset of cases. Until recently the focus of debate about the formal listing of houses and neighbourhoods was on pre-twentieth century structures, but now, for various reasons, a number of more modern buildings have been identified as qualifying for protection. Included within this new category are examples of blocks and estates originally built by local authorities ('council housing') but now, following the impact of the right to buy since 1980, mixed tenure neighbourhoods in multiple-ownership. This may not, in itself, be a cause for concern, but other factors such as the changing character of social housing, especially the declining popularity of some estates, can lead to the threat of radical remodelling or even demolition. In other cases it is not unpopularity but the ageing process that prompts intervention. Later sections consider two quite different examples of council built estates, each of which has a claim to be considered a valued historic environment under threat. The Sea Mills estate in Bristol is not formally listed but it enjoys conservation area status, reflecting its high quality 1920s garden suburb layout. It is currently also subject to plans that will lead to partial redevelopment. The Byker estate in Newcastle upon Tyne is an inner city area built in the 1970s, which was threatened with redevelopment but is now listed on account of its architectural merit.

Valuing Housing

In physical and architectural terms houses are quite simple structures, even when they are part of multi-dwelling blocks. But a broader approach suggests that houses are actually much more complicated and multi-dimensional.[1] As a consequence, there are different ways of attaching and measuring value in houses. The sorts of attributes that people value in their homes would include (in no particular order): aesthetics, location, utility, asset worth and sentimental attachment. Location is important in many ways, not least to the extent that it influences access to scarce spatially distributed resources, such as schools and job opportunities. A home is also an address, which says a lot about the occupants' social and economic status. Utility covers the fact that a house provides shelter from the elements, a store for possessions, an arena for social interaction and sometimes a site of economic activity. The home that is valued is one that provides a degree of security and privacy for family life. John Turner[2] long ago argued that housing is a verb: dwelling needs to be recognized as an activity as well as a structure, and it is reasonable to suggest that what people value is the extent to which their house enables them to make a home in it (or of it). However, increasingly the attractiveness of a dwelling lies not just in the housing services that it provides (its homeliness) but also in its ability to act an asset that can be made to deliver other benefits.[3]

In contemporary Britain 70 per cent of homes are owned by their occupiers, and this proportion is expected to rise.[4] Private ownership of property is a fundamental value of market based societies in general and freedom to maintain, modify and improve one's home is presented as one of the advantages of owner occupation. Many people like to stamp their personal mark on their home – this was particularly noticeable in the years after 1980, when council tenants were given the right to buy their homes, as purchasers adopted a variety of techniques for proclaiming that they were no longer tenants of the council. However, the freedom of property owners is not unconstrained: there is a tension between their rights and freedoms and the legitimate interests of the wider community (including posterity) represented by the planning system. The law requires that planning permission is obtained for both new development and for certain sorts of modifications to all existing properties. Thus the rules and guidelines relating to listed buildings of special architectural or historical interest need to be seen as an extension of the planning system that applies to the built environment as a whole. They constitute an additional set of detailed restrictions with which property owners need to comply.

---

1   P. King, *Private Dwelling*, (Aldershot: Ashgate, 2006); W. Gallagher, *House Thinking. A Room by Room Look at How We Live*, (New York: Harper Collins, 2006); B.M. Lane, (ed.), *Housing and Dwelling: Perspectives on Modern Domestic Architecture*, (London: Routledge, 2006).

2   J.F.C. Turner, and R. Fichter, (eds), *Freedom to Build*, (New York: Macmillan, 1972).

3   P. Malpass, 'Housing and the new Welfare State: Wobbly pillar or cornerstone?', *Housing Studies*, 23, 1, (2008): 1–19.

4   S. Wilcox, *UK Housing Review 2007/2008*, (Coventry and London: Chartered Institute of Housing and Building Societies Association, 2007).

Valuing housing as heritage involves a rather different set of factors from those identified in relation to consumers. Planning Policy Guidance note 15 (PPG 15)[5] set out the following criteria for listing:

- architectural interest,
- historical interest,
- close historical association with nationally important people and
- group value.

These have been updated in a new Circular from the CLG (CLG Circular 01/2007),[6] which emphasizes architectural and historic interest, but also mentions age and rarity and aesthetic merit. Close association with important people seems to have been dropped.

PPG15 lists in detail the sorts of things that can and cannot be done to the internal and external fabric of listed buildings. This is mentioned here to highlight the tension between the rights and freedoms of property owners and the claims made on behalf of posterity.

Housing as 'Heritage' and as 'Heritage'

As a simple analytical device this chapter employs a distinction between 'Heritage' and 'heritage'. At one end of the spectrum lies Heritage as the outright commodification of the past, where the objective is to make money by charging people for the opportunity to visit a site of particular historical interest, be it a castle, stately home, 'lost' garden or ruined abbey. In the majority of such cases there is no-one living there, or if there is their quarters usually remain private and only tantalizingly visible. In this context the value of the property lies in its intrinsic historical significance, but also in what people will pay to visit it, not in its utility as a dwelling or whatever it was built as. Further along the spectrum is a different sort of Heritage, where the past is being used as part of a place marketing strategy, but where visitors are not charged for entry and where a multitude of people both live and work. Places such as Bath and Stratford-upon-Avon would be good examples. The difference between the two cases is that in the former the Heritage site is the primary attraction, and it is primarily an exhibit. In the second case the Heritage site is a place where people live and work, many of them in jobs that are unrelated to the presence of Heritage. As residents their relationship with the Heritage is different from that of visitors who are there mainly to look, have their photograph taken and possibly learn something interesting. In this case the

---

5   Department of the Environment and Department of National Heritage, *Planning Policy Guidance 15: Planning and the Historic Environment*, (London, HMSO, 1994).

6   Department for Communities and Local Government, *Revisions to Principles of Selection for Listed Buildings, Circular 01/2007*, (London, TSO, 2007).

buildings have to work as homes or places of employment as well as exhibits, demands which may not be easily reconciled.

Further along again there is Heritage in the form of buildings in private ownership and use, but which are nevertheless publicly and formally identified and listed. Here the buildings are not primarily exhibits and the cash nexus is absent or much less intrusive; indeed it can be argued that what is important is the protection of valued historic buildings from the predations of unsentimental market forces. The justification for listing in these cases is the intrinsic merits or historical significance of the building, irrespective of whether anyone ever comes to look at it or tries to make money out of visitors.

Finally, there is heritage, in the dictionary sense of that which is inherited. This can be taken very broadly, to include everything, all material structures and objects and cultural practices of all kinds. In the case of housing the heritage consists of all 26 million dwellings in the United Kingdom. Obviously the great majority of these have no great architectural merit or historical significance, nor are they associated with notable people or events. They are simply, but importantly, the structures in which people make their homes, nurture their children and conduct family life. As such they are not Heritage, but they are nonetheless a valuable form of plastic heritage, durable yet adapting to the demands of a rapidly changing society.

This chapter focuses mainly on housing as heritage, and in particular the recent idea that council-built housing estates can be treated as Heritage. It looks at the interface between heritage and Heritage, exploring some of the issues raised by giving special status to places that are also people's homes. It is about identifying and discussing the tensions, not resolving them.

The Importance of Housing as Heritage, the Implications of Heritage

The major urban land use in Britain is residential: it is houses, both large and small, cottages and blocks of flats, set out in streets, avenues and squares, which give the country's human settlements their distinctive character. To a large extent they *are* the urban landscape. The housing market is important at different scales: at the local level it is what determines the social character of neighbourhoods (similar houses in different housing markets can have very different social characteristics). On the larger scale the housing market is acknowledged as an important component of the economy and a key macro-economic indicator. What is happening in the housing market can influence levels of consumer spending, with implications for employment and investment. The housing system itself provides large numbers of jobs, in a variety of different settings. For present purposes it is sufficient to refer mainly to construction and maintenance, but there is also the do-it-yourself industry. All of these have direct impact on the nature and quality of the housing stock as a whole.

Great Britain has a large but rather old, and ageing, housing stock. In 2004 39 per cent of homes were over 60 years old,[7] and because of low levels of new building in recent years the average age has been increasing. At current rates of replacement each new house needs to last for an improbably long time. Some years ago the figure for England was calculated to be 3,300 years, reflecting very low rates of replacement.[8] In this situation we cannot afford not to regard every dwelling as a valuable resource for the indefinite future. Home owners behave as if they understand this, actively investing in repairing and modernizing their homes. Up to date figures are not readily available, but an estimate for repair and decoration expenditure by home owners (but not including improvement) implied a grand total of nearly £9 billion being spent in 1996/97. Despite this impressive level of spending, 6 million homes in England (27 per cent of the total) are deemed to be below the 'decency standard'.[9]

Precisely because every house has to be made to last, and to be habitable, for a very long time into the future, we cannot afford to say that any houses should be put on a pedestal and insulated from change. Flexibility and utility must rule. In this situation, therefore, to designate any of the valuable housing heritage as Heritage is to risk imposing restrictions on the capacity of people to make homes for themselves in ways and in conditions that accord with contemporary standards and expectations.

A critique of the idea of housing as Heritage suggests that it strikes the wrong note and misses almost everything that is important about housing. First, it is elitist, in two senses: it focuses attention on a tiny minority of (allegedly) historically significant dwellings or blocks of dwellings, and it picks them out for special treatment on grounds specified by particular professional interests, dominated by design aesthetics, grounds that are largely unrelated to, and potentially in conflict with, the reasons that the people living there value the properties. Second, Heritage is inherently backwards-looking, invoking history as the basis of listing. For very old buildings age alone is sufficient to warrant listing. Third, it is also inherently negative: as mentioned above, it is about setting out what cannot be done, imposing additional constraints on the extent and nature of change beyond those of normal planning policy. Fourth, Heritage elevates property both over the people who have to live in and with it, and over the processes that produce buildings. Listing tends to place heavy emphasis on the building itself, as an object of importance in itself, abstracted from the context in which it was created and separated from the people who use and interact with it.

---

7   S. Wilcox, *UK Housing Review 2006/2007*, (Coventry: CIH and London: CML, 2006), 127.

8   P. Leather and S. Mackintosh, *The Future of Housing Renewal Policy*, (Bristol: SAUS Publications (University of Bristol, 1994), 4.

9   Department of Communities and Local Government, *English House Condition Survey 2005 – Headline Report*, (London: DCLG, 2007).

In contrast to the Heritage approach, seeing housing as heritage is non-elitist, emphasizing instead the value of all dwellings. Second, it is a forward looking and positive approach, concerned about the need to keep all houses in use indefinitely. As such it focuses on the advantages and benefits of modernization and improvement. It is about ensuring the continued utility of the building rather than the architectural purity and integrity. Third, it is underpinned by a different stance on the relationship between houses and the people who live in them, recognizing that dwelling is a verb as well as a noun. The problem is to achieve a reasonable balance between these two parts of speech.

Housing at the Heritage/heritage Interface

Housing, especially occupied housing, poses particular problems for the Heritage lobby because of the potential tension between the wish to retain the integrity of the structure and the owner's wish to make changes. The argument developed above might be read as implying that the needs and preferences of owners and users should over-ride the claims of Heritage. Buildings in general, including great architecture, inevitably reflect society, not *vice versa*. Buildings only exist because they serve a social or economic purpose, which may not be the one originally envisaged. This is particularly applicable to ordinary houses, and is the basis for arguing for the subordination of the built environment to social change. As society changes so buildings must be flexed to reflect new needs and demands. As long as people need to make their homes in old buildings then they must be allowed to do so with reasonable freedom and in reasonable comfort. Once buildings cease to serve any useful purpose and are incapable of being adapted at reasonable cost then surely they must either go, or become mere exhibits, transformed into tourist attractions.

However, fortunately this is not a case of either/or: we can preserve both the architectural and aesthetic character of neighbourhoods and have housing that keeps up with changing expectations. There are huge numbers of late Victorian and Edwardian houses that are highly valued as homes in the 21st century, but no-one would want to live in them as built, with rudimentary plumbing, no central heating, hot water or inside toilets, and completely reliant on fossil fuel for cooking and warmth. The great majority of these same houses were built without access to mains electricity. It is also probably true that few people would be happy to live in 1960s houses in their original condition, as the success of the replacement window industry testifies. The Government's Decent Homes Standard (DHS) specifies that to meet the standard houses need to be in a reasonable state of repair and have reasonably modern facilities and services (this means a kitchen less than twenty years old and a bathroom less than thirty years old).[10] Decent homes are also

---

10　Department of Communities and Local Government, *A Decent Home: Definition and Guidance for Implementation*, (London: DCLG, 2006).

required to have a reasonable degree of thermal comfort, provided by an efficient heating system, something notably lacking from the great majority of new houses built in Britain until the latter part of the last century.

The Poverty of Policy

It is not technically difficult to install modern facilities and services in old buildings, keeping them in use for prolonged periods, and at the same time preserving the character of the neighbourhood. However, it can be financially difficult, especially in the context of the artificially constructed constraints of local government finance. The Sea Mills estate in Bristol, for example, is now facing extensive redevelopment over the period up to 2013 because of a complex set of legal, political and financial factors which make retention and repair of the existing defective houses unfeasible. The result will be increased housing densities despite restrictive covenants specifying that the land must be used for garden suburb accommodation, and despite parts of the estate enjoying conservation area status.

The Sea Mills estate, built between 1919 and 1931, was one of the first areas of council housing developed in Bristol. As such it benefited from the rhetoric around 'homes fit for heroes' in the immediate aftermath of the Great War, being built to a high standard (compared with what came later) and laid out on garden suburb lines featuring low residential density, wide tree-lined roads with grassy verges and generous landscaping (see Figure 10.1). Sea Mills also benefited from the fact that it acquired, and retained, a reputation as a good, socially desirable neighbourhood, which, in the long run, helped to insulate it from the impact of residualization. Bristol City Council (BCC) recognized the quality and charm of Sea Mills by according the southern part of the estate conservation area status in 1981. The estate is to this day a fine example of early public housing and a tribute to what can (or could) be achieved by properly resourced local authorities. This was reflected in the planning brief for the redevelopment: 'The form and layout of the estate, the picturesque countryside backdrop, relatively low density and large gardens create a spacious character and appearance'.[11]

However, some of the individual houses are now nearly 90 years old and although they have been modernized over the years it has been determined that 250 houses (known as Parkinson-type houses), built using a particular system of concrete panel construction, have come to the end of their useful life. The Parkinson houses have become structurally unsound due to corrosion of the steel frame supporting the concrete panels. Some (those that had been bought under the right to buy in the early 1980s) were comprehensively repaired in the 1980s and 1990s by replacing the concrete cladding with brickwork. This looked dramatic, as the roof was supported on jacks while the walls were removed and rebuilt. It was

---

11 *PRC Redevelopment Project, Sea Mills Planning Brief, May 2008*, paragraph 8.28, http://www.bristol.gov.uk/ccm/cms-service/stream/asset/?asset_id=27542004.

Figure 10.1   The Sea Mills estate, Bristol
Source: © Peter Malpass

also expensive, but it did have the advantage of preserving the general appearance of the estate, and of course it made life much better for the owners. In some cases they were also provided with new kitchens and bathrooms at the same time as the structural repairs. The majority of privately owned defective Parkinson houses at Sea Mills were in fact completely rebuilt on the original footprint. The Council had a statutory duty to carry out the repairs for private owners, but it was not obliged to do the same for its own identical houses, and neither did it have the resources to do so. However, where a privately owned house was paired with a Council owned house then they were both remedied.

In 2000 the Government introduced the Decent Homes Standard, mentioned above, requiring all local authorities to bring their housing stock up to the standard by 2010. This was followed by a requirement to devise a 30 year business plan for the Council housing stock, and to carry out a comprehensive options appraisal pending the adoption of policies to ensure that the business plan, including meeting the DHS, could be achieved. The subplot here was that the Government's strategy was designed to bring local authorities face to face with the financial implications of continuing as housing landlords. There was the clearest possible encouragement to them to opt for transfer of ownership to a registered social landlord, which would ensure access to the financial resources necessary to meet the costs of the DHS. However, in Bristol (and nearly 100 other local authorities) the outcome of the options appraisal was a decision to retain ownership of the stock. In order for this decision to be signed off by the Department for Communities and Local Government it needed to be both financially viable, and receive the support of tenants. BCC was able to demonstrate, at the time, that the business plan balanced with stock retention (albeit with some revenue savings, and alternative options for the worst stock).

Sea Mills is, therefore, to some extent a victim of circumstances, in particular the tension between central Government policy and local preference. A Government pulling in one direction, seeking to use financial leverage to effect the demunicipalization of social housing, created difficulties for a City Council that was inclined in another direction, and that was required to make financial adjustments in order to hang onto its housing stock. The consequence is that the charm and design integrity of the estate are threatened by the Council having to adopt policies that not only involve the demolition of the defective houses but redevelopment in a way that significantly increases the overall density. There are currently 249 3-bed precast reinforced concrete (PRC) properties, and the initial development proposals (IDPs) show these replaced with a total of 373 houses and flats (including two backland sites, and a further three additional parcels). Importantly, there are 164 house-type flats within the IDPs (in place of 82 PRC houses). The Council's legal advice is that the proposals are within the original spirit of the covenants specifying use as a garden suburb, covering arrangement of dwellings, distance between homes, gardens and so forth. Initial feedback from the City's Planning Department and English Heritage indicates that the proposals for the layout of the new homes are within the constraints of the Conservation Area. It is even considered possible that the design of the new homes could enhance the conservation area creating a modern garden suburb.

Nevertheless, despite the optimism within the local authority, it is not in full control, for in order to complete the work it needs to involve a private developer, and not simply as a contractor, as in the past when the estate was built. BCC will have complete control over the mix of types, size, specification, location of the replacement council homes, although not the actual number, which will reflect the overall economics of the project from the developer's perspective. The developer will have a degree of freedom over the private units, within the confines of the statutory planning system. The plan is to come to an agreement with a developer such that within the terms of the planning brief, the developer will have a degree of freedom to decide what sort of dwellings to build, and how many. In return for this marketing opportunity, on land transferred from the Council, the developer will deliver an agreed number of council houses back to the local authority. It is accepted already that the number of council houses will be smaller than at present, even though the total number of houses on the estate will increase (the balance between open-market houses and council houses might be two to one). The problem here, from the point of view of the heritage value of Sea Mills, is that the more restrictions the Council seeks to impose on the developer the more the value of the land falls and the fewer council houses they secure from the deal.

This is a situation in which both sides will be seeking to secure maximum value for themselves, but defined in different ways. For the developer it will be the mix of dwellings that produces the greatest return in the market place, while for the Council it is the largest number of council houses obtainable for minimum concessions in terms of dilution of the special character of the area.

It is important to acknowledge the point made in the draft planning brief, that the individual houses have no particular architectural merit – it is the overall effect of the way they are grouped and laid out that creates the character and appealing appearance of the area. The problem is that although the estate is designated as a conservation area the local authority lacks the financial resources to rebuild the houses *in situ* in a way that would retain the character and appearance of the area. Politics and economics trump the environment. Different approaches were considered; for example, in theory, given that the local authority already owned the land it could have commissioned a contractor to build, to the Council's specification, a certain number of houses which could then have been sold at affordable prices (because the land value was retained by the Council). But this sort of approach was ruled out, partly by local insistence on providing rented council owned houses, and by the economics: in order for the replacement of the defective council houses to be achievable it was necessary to generate resources through the development of additional houses that would be sold on the open market. It should also be acknowledged here that even if the Council had been in a position to undertake the work itself, it might still have opted for a programme of planned demolition and rebuilding, with increased densities. Arguments for this sort of approach would include references to the small size of the existing houses and the case for incorporating lifetime home standards and improved thermal efficiency which are easier to achieve in new buildings. There is also a case for providing a different mix of dwelling sizes in an area dominated by three bedroom houses.

It is, of course, too early to speak confidently about how the Sea Mills project will turn out, but a provisional conclusion would be that the Council has arrived at a position (through a combination of Government policy and its own way of responding) where there is more risk of the distinctive qualities of the estate being diluted than would have been the case had the Council been able to deal with the problem itself. Current housing policy at national level, dominated by a strong preference for demunicipalization and market based solutions, is insensitive to the claims of valued historic neighbourhoods like Sea Mills.

Material Heritage and Cultural Practices

Turning to a rather different sort of problem, while it is possible to argue that great buildings are themselves cultural artefacts that should be preserved, it is also arguable that buildings are to be understood as the products of cultural processes, and to preserve individual buildings in rapidly changing societies is to imply that the building is somehow more important and worthy of preservation than the

Figure 10.2   Byker in 1975, showing the contrast between the old, on the right, and the new, on the left
Source: © Peter Malpass

social and economic institutions and arrangements that called it into existence. Some people would see this as an inverted set of priorities. The campaign group Defend Council Housing,[12] for example, would want to protect the principle of municipal provision of decent affordable housing as a valuable social and political achievement of the twentieth century which is in danger of becoming extinct in the twenty first. On this view, council housing as an institution is more valuable than particular houses or estates. In a sense listed former council estates are relics of a system of housing provision that has been systematically wrecked in the last thirty years or so.

A related argument is that to preserve buildings alone is to run the risk of losing sight of their true meaning or significance. Preserving the buildings is actually easier than preserving and communicating the story behind them. Take, for example, the case of the Sea Mills estate discussed above. To understand estates like Sea Mills it is necessary to contrast it with residential neighbourhoods laid out only a decade earlier. Before 1914 garden suburb standards were too expensive for the working class, but in the aftermath of the Great War it was necessary to demonstrate that nothing but the best would do for the heroes who had won the war.[13]

---

12   www.defendcouncilhousing.org.uk.
13   M. Swenarton, *Homes Fit For Heroes*, (London: Heinemann, 1981).

The Byker estate in Newcastle tells a different story. Without in any sense denying the quality of the architecture, the real significance of Byker lies in other aspects of the story. First, the fact that the Byker estate was designed by a top international architect, Ralph Erskine (and the team based in Erskine's Byker office), tells us something about the role and status of council housing in the 1960s. Historically architects have been associated with building for the rich and powerful, producing pyramids, castles, palaces, cathedrals and the like. The involvement of leading architects in the production of working class housing was largely a product of post-Second World War reconstruction in a particular socio-political context, known as the welfare state. It was also part of the wider modernist project to sweep away the past and to create a shiny new world. Places like the old Byker were swept away because they were old, and it is a nice irony that the replacement buildings are now listed.

The contrast between then and now is instructive. In the 1960s, when Ralph Erskine was commissioned to work on Byker, council housing was a force in the land, with up to 200,000 new dwellings being built each year. Council housing was an integral part of the modernization of British cities, at the cutting edge of urban renewal. Erskine's plan involved 2,400 dwellings, a scale of investment that is completely inconceivable as a social housing project today. As a result of the changes in policy since those days leading architects are no longer working on social housing. The very small scale of most current projects is one factor, but also reliance on procuring social housing on the back of private sector development means that the design is usually entrusted to the in-house design teams employed by the volume house builders. Of course, there is an argument to be made to the effect that most council housing was not produced by top designers, and that the entanglement of architects with mass housing brought more disasters than triumphs,[14] but the new private estates currently being built suggest that market forces cannot be relied upon to produce good design either.

The reason that Erskine was appointed to rebuild Byker was that the City Council had responded to demands from the people of the neighbourhood who were demanding that the redevelopment should be done in a way that would preserve valued aspects of the existing community. A key element of this came down to offering local rehousing to those who wanted to stay, and Erskine's team devised a phased programme of clearance and rebuilding to enable this to happen, at least to some extent. In practice the goal of retaining the community faded from view after 1970 as priority was given to overcoming the practical difficulties of getting house built. Forty per cent of new homes were completed after the last of the old ones had been demolished and 5,000 or so households left the area with no guarantee of return.[15] Erskine's architectural achievement in Byker is rightly acknowledged,

---

14   A. Coleman, *Utopia on Trial*, (London: Hilary Shipman, 1985), see also N. Teymour, T. Markus, and T. Woolley, (eds), *Rehumanising Housing*, (London: Butterworths, 1988).

15   P. Malpass and A. Murie, *Housing Policy and Practice*, (Basingstoke: Macmillan, 1987), 221–2.

but it is arguable that the social project of retaining the community has been much less successful, with part of the estate eventually coming under threat of partial demolition as a result of social problems.[16] This might be partly because it was always based on flawed social theory, partly because the old community was already under threat from social and economic changes that were incapable of being transcended by any designer, and partly because Byker as a community has suffered the ravages of housing policy since 1979 as council housing has been denigrated and undermined. The point, however, is to note the irony of the way that Byker was rebuilt to save the community, but now it is the built form that is listed for preservation, despite the failure of the original plan to save the community.

It is arguable that to list social housing estates today is as mis-guided as was the idea that it was possible to preserve the community through devices such as local rehousing, 'streets in the sky' or any of the other tropes of socially aware designers in the 1960s. Listing the physical fabric is an inadequate response to the rising tide of residualization that is sweeping through social housing, driven in part by polices that continue to laud home ownership over other forms of consumption. As long as social housing is seen as essentially a safety net for the least well off in society then it will be difficult to prevent further decline. Indeed it is even possible that the restrictions resulting from listing will actually make it more difficult (and more expensive) for local authorities to produce homes and environments that people want to live in.

Conclusion

To conclude, this chapter has sought to make five main points: first, that a focus on housing as Heritage misses most of what is important about housing; second, that all housing is a form of heritage to be valued and cherished for the indefinite future, and therefore, flexibility and utility must take priority; third, that it is possible to reconcile the need to protect existing residential neighbourhoods and at the same time allow people to live in homes that respond to changing needs and preferences; fourth, current policy, as illustrated by the case of Sea Mills, makes it very difficult for local authorities to implement sensitive policies designed to retain the character of valued neighbourhoods, and fifth, the listing of buildings elevates the importance of design and physical structure over the cultural processes and achievements of which they are a reflection.

In the present era of concern about the environment and the impact of human activity on climate change it can be argued that it is equally wrong in principle both to list some buildings to save them from demolition and to demolish *any* building that has (or is capable of being given) continuing utility. It is in the interests of both the users of buildings and the planet as a whole that utility rules. There might

---

16   See this volume: J. Pendlebury, T. Townshend, and R. Gilroy, 'Social Housing as Heritage: The Case of Byker, Newcastle upon Tyne'.

be a case for listing some buildings that have no ongoing use value other than as exhibits, but there should also be a general presumption against demolition and in favour of measures that keep buildings in use.

Bibliography

Coleman, A. *Utopia on Trial*, London: Hilary Shipman, 1985.

Department of Communities and Local Government. *A Decent Home: definition and guidance for implementation*. London: DCLG, 2006.

Department of Communities and Local Government. *English House Condition Survey 2005 – headline report*. London: DCLG, 2007.

Department of Communities and Local Government. *Revisions to Principles of Selection for Listed Buildings, Circular 01/2007*. London, TSO, 2007.

Department of the Environment and Department of National Heritage. *Planning Policy Guidance 15: Planning and the Historic Environment*. London, HMSO, 1994.

Gallagher, W. *House Thinking. A Room by Room Look at How We Live*. New York: Harper Collins, 2006.

King, P. *Private Dwelling*. Aldershot: Ashgate, 2006.

Lane, B.M. ed. *Housing and Dwelling: Perspectives on Modern Domestic Architecture*. London: Routledge, 2006.

Leather, P. and Mackintosh, S. *The Future of Housing Renewal Policy*. Bristol: SAUS Publications (University of Bristol), 1994.

Malpass, P. 'Housing and the new Welfare State: Wobbly pillar or cornerstone?', *Housing Studies*, 23, 1, (2008): 1–19.

Malpass, P. and Murie, A. *Housing Policy and Practice*. Basingstoke: Macmillan, 1987.

Pendlebury, J., Townshend, T. and Gilroy, R. 'Social housing as heritage: The case of Byker, Newcastle upon Tyne'. In *Valuing Historic Environments* edited by Lisanne Gibson and John Pendlebury, 179–200. Farnham, Ashgate, 2009.

PRC Redevelopment Project, Sea Mills Planning Brief, May 2008, http://www.bristol.gov.uk/ccm/cms-service/stream/asset/?asset_id=27542004.

Swenarton, M. *Homes Fit For Heroes*. London: Heinemann, 1981.

Teymour, N., Markus, T. and Woolley, T. (eds) *Rehumanising Housing*. London: Butterworths, 1988.

Turner, J. F. C. and Fichter, R. (eds) *Freedom to Build*. New York: Macmillan, 1972.

Wilcox, S. *UK Housing Review 2006/2007*. Coventry: CIH and London: CML, 2006.

Index

For Product Safety Concerns and Information please contact our
EU representative GPSR@taylorandfrancis.com Taylor & Francis
Verlag GmbH, Kaufingerstraße 24, 80331 München, Germany